DEVELO
BIODYNAMIC
AGRICULTURE

Reflections on Early Research

Adalbert Count Keyserlingk

English by A.R. Meuss, FIL, MTA

TEMPLE LODGE
London

Temple Lodge Publishing
51 Queen Caroline Street
London W6 9QL

www.templelodge.com

First English edition by Temple Lodge, 1999

Originally published in German under the title *Erinnerungen an Frühe Forschungsarbeiten* by Verlag der Kooperative, Dürnau, Germany 1993

© Verlag der Kooperative 1993
This translation © Temple Lodge Publishing 1999

The moral right of the author has been asserted under the Copyright, Designs and Patents Act, 1988

All rights reserved. No part of this publication may be reproduced, stored in a retrieval system, or transmitted, in any form or by any means, electronic, mechanical, photocopying or otherwise, without the prior permission of the publishers

A catalogue record for this book is available from the British Library

ISBN 1 902636 13 9

Cover painting by Olive Whicher. Cover layout by Clairview Studios
Typeset by DP Photosetting, Aylesbury, Bucks
Printed and bound in Great Britain by Cromwell Press Limited, Trowbridge, Wilts.

Contents

Publisher's Note	iv
Foreword *by Manfred Klett*	1
Insights gained from early work in biodynamic research	3
Sasterhausen estate	13
Differences between East and West in the earth organism	20
Biodynamic preparations	32
Stag's bladder and yarrow	32
Stinging nettle	33
Peritoneum and dandelion	34
Ashing	38
Getting to know the pioneer plant breeders	40
Transforming plant species and varieties *by Immanuel Voegele*	51
Koberwitz remembered	63
Learning about scientific research at the first Waldorf school in Stuttgart	76
Some correspondence with Miss Windeck	80
Photographs of the experiments done at Pilgramshain	89
The revitalization of the ether powers *by Erich Kirchner*	100
Notes and References	106

Publisher's Note

Following our publication of *The Birth of a New Agriculture* (edited by Count Keyserlingk), we are pleased to present this second volume which enters into details about the development of biodynamics and, in particular, research conducted after Rudolf Steiner's death.

As with the previous volume, this book is primarily intended for those who have an interest in Rudolf Steiner's spiritual philosophy, Anthroposophy. Specialist terminology and concepts are referred to throughout the book, often without qualification.

Count Keyserlingk died in October 1993. In January of that year he wrote (see page 88) that the manuscript for this book had not 'turned out the way I intended'. Evidently he would have liked to have done further work on the book. It is understandable, then, that the German edition contains some inconsistencies and a few inaccuracies. These have been corrected by the translator, who has also added some references. Some further minimal editing has also been undertaken. However, as this is very much Count Keyserlingk's 'last testament', we felt that a general rearranging of the text would not be appropriate—despite the sometimes unconventional structure of the book.

We would like to thank Karin Herms who worked hard to ensure publication of these two important volumes in English translation. We would also like to thank Countess Keyserlingk for giving us permission to publish, and John Brett and the Camphill Communities of Thornage Hall and Botton Village for their kind sponsorship.

SG, London, September 1999

Foreword

From the vantage point of his great age, Adalbert Count Keyserlingk, who died on 23 October 1993, was looking back, above all remembering the early days when the impulses of Rudolf Steiner's agriculture course, given at the Count's family home, Koberwitz Manor near Breslau (Wroclaw) in Silesia, were beginning to spread. His mind went back to many a pioneer who sowed the seeds of a new agriculture in this century, in spite of all obstacles. Farmer and physician, he remembered the many efforts he himself had made at the Sasterhausen estate in Silesia to explore Rudolf Steiner's suggestions in his experiments and put them into practice. Everything came from a dual motif in his biography—to develop the medical impulse that came from the source spring of anthroposophy and to make it bear fruit in the field of plant breeding, or, to use the preferred term, plant transformation. It is this, no doubt, which makes the book important in our present time. Many people worked hard in those early days to gain a personal relationship, founded in the spirit, to cultivated plants; in outer terms by doing experiments, often original, and in inner terms by means of meditation. Taking up the suggestions made by Rudolf Steiner they also sought to breed cultivars from wild plants. Many of those efforts may have remained germinal, but is it right to think that we can go our own way without first taking note of them? Just as it is part of biodynamic agriculture to consider developmental aspects in all one's work, so the ideas or will impulses of all those who have been involved in the process of evolution need to be properly appreciated to give this approach to agriculture its continuity in the spirit.

Adalbert Count Keyserlingk's book ranges widely, from the spiritual roots which he found in the School of Chartres to his personal experience of Rudolf Steiner's work at the Stuttgart Waldorf School and in Breslau and Koberwitz, his reflections on the work that was done to apply the agriculture course in practice and how he developed his own research questions from it. This is an authentic description of some of the history of the biodynamic movement, reflecting the enthusiasm, the exalted feelings and the power of initiative that came to people from Rudolf Steiner's work at Whitsun 1924, the tragic history of Koberwitz estate and of his father, who was frustrated in his desire to provide a sound basis for the new impulse, the loss of all the Count's own research records during the war, and finally his lively interest, right into the present time, in everything which those who came later have contributed to the further development of the biodynamic movement.

Manfred Klett
Dornach, May 1993

Insights gained from early work in biodynamic research

During the agriculture course[1] given at Whitsun 1924 and in my parent's home in the years that followed, the course was set for my life at a time when I was 19 years of age. Now, as an old man, I'd like to tell of the work done by myself and a small group of farmers before the Second World War, as we sought to maintain and renew the food plants that will be our staff of life in future times.

Let me begin with the words of the great Alanus ab Insulis, Master of Chartres, who gave the impulses for bringing the Christian spirit into the 'seven free arts' and made people see the goddess Natura as a spirit active in the world all around them. His words sustained the development of Western humanity and still shape our lives today wherever people make it their heart's desire to maintain and breed the plants that feed us. It can be a help in our work to take these words into our hearts and minds. The suggestion to study the work of Alanus ab Insulis came from my friend Wilhelm Rath and his translation of the *Anticlaudian*, where he writes the following.[2]

> It was his great concern for the future of mankind that made Alanus put the following words in the mouth of the goddess who appeared before him:
>
> How soon, alas, does a step take us away from virtue
> And let it fall victim to vice.
> There virtue loses all her beauty
> And madness loosens the bridle kept on vice.
> Alas, how the light of the Sun's justice is fading!
> All that remains are shades of the shadow.
> She weeps for her heavenly light, for it has gone,

> Night now flows over her, lightless as she mourns.
> Only the lightning flash of evil deed now shines on this earth,
> Night of delusion clouds our faith.
> No redemption comes from light of virtue's star
> To bring liberation from this abyss!
> Twilight of faith covers the world
> And the chaos of delusion weighs on us all.

Wilhelm Rath then goes on:

> But the shadow points to something that is of the essence, something from which the shadow comes. Our thoughts thus make us aware of spiritual realities and we can find the way to these within us if we are ready to set out on it with courage and a pure heart. That was the deeper meaning of the whole of his life's work, presented in poetic form in his *Anticlaudian*.

Much can be said and has been said about the shadow, both in Samothrace and among the Celts and above all in the mysteries of Megalithic civilization which I have described in my book *Und sie erstarrten in Stein*.[3]

The shadow was felt to be the image and soul of the being who casts it. It comes with the sunrise, being there suddenly, as though coming out of infinity. Stone Age people would lie in the shadows of the menhirs to dream of the one who cast the shadow, of his nature and work. Insights were thus gained in the 'shadow mysteries'.

The great experience people had during the temple sleep of the mysteries, lying by menhirs and stone circles, was that they discovered the true essence of the one who cast the shadow. Every blade of grass, every tree, human being or animal always casts a shadow in a particular direction, above all at sunrise and sunset. We should learn to pay attention to this again.

Alanus ab Insulis wrote of such a shadow experience in his *Apocalypsis:*

From the direction of Taurus the star of Apollo sent searing heat
Pouring red-hot arrows on my head and my brain.
Then I entered the protective shade of a grove—
To gain the grace of cooling Zephyr's kindness.

The heat of noon seemed like summer to me,
But under the leafy canopy I lay safely protected—
When Pythagoras's form came to me, very close—
If it was he embodied—God knows I know not.

And as I look upon the image of the sage
I see it filled with many a sign and script.—
Am I out of my body as I see his face,
Or in my body—God knows I know it not.

From his brow the art of astronomy shines forth,
Grammar brought order and harmony to his teeth,
Rhetoric flowered in beauty on his tongue,
Whilst logic flowed on his eager lip.

To the fingers belongs the art of arithmetic,
From his throat streams music's world of sound,
In his eyes, geometry shows itself full of care,
Each part of the body has its art in him.

On his breast I find the foundation of all ethics,
The skills of artisans are writ on his back.
As I sought to read further in the body's book
He himself opened it, uttering the words: 'Look here!'

And showed me the secret of his right hand,
I looked and read what was written there.
In dark signs a script here showed itself:
'I want to be your guide, and you shall follow me!'

Wilhelm Rath continued:

The deep connection Alanus had to the nature of the seven free arts, which people saw in him in his day, and the fact that they called him a 'Father', a seer, can also be understood when we read these words of his.

When Rudolf Steiner began to speak of the School of Chartres and its great teachers, he said of Alanus, and I am giving his words in full:

> And above all there was one man who really was greater than any of them and who, if I may put it like this, worked out of an ideal inspiration as he taught the seven free arts in their relationship to Christianity. This was Alanus of Lille who fired up his students with his teaching in Chartres during the twelfth century. He truly realized that in the centuries that lay ahead teachings offered in this way would no longer benefit the earth. For this was not just Platonism, it was the teaching that came from vision gained in the mysteries in pre-Platonic times, only that the vision had also taken Christianity into itself.

We then read how Alanus explained the seven free arts and we get the picture of a carriage being built by them:

> The arts build a carriage for Intelligence. Grammar provides the shafts, logic the axle, rhetoric adorns the shafts, the fourth of them, arithmetic, gives the first wheel, music the second, geometry the third and astronomy the fourth. Intelligence then harnesses five horses, the five senses—sense of sight, sense of hearing, sense of smell, sense of taste and sense of touch. Common Sense is the coachman and Intelligence rides in the carriage.

We must seek to make connection with the seven free arts, which in those times were still very much felt to be spiritual entities; we must seek to come closer to the goddess Natura with faithful and consistent work and practice. The goddess unites within herself all the elemental spirits and group souls of which we speak, and we must ask them to serve with us, to help us in our work, when we are involved with her 'children', the plants. The aim of all plant breeding work is to let a transformed plant, a new child, arise that will be able to give human beings new vital energies.

Just as we human beings are able to produce children and future generations, living beings—a capacity shared by all of nature, all animals, all plants—so the elemental spirits and group souls are able to create something new, let something new arise in the world, so that the world, too, may have new life. These processes, which then also take physical form, are important to every plant or animal breeder, who needs to live in and work with the two worlds that are important for the life of nature. And this is something anyone can do who seeks to come closer to creation by gaining insight through the heart.

In the above-mentioned book I was able to show that at an early stage in Earth evolution the planetary spirits also joined forces to create new things. Examples are Venus uniting with the earth so that copper arose, or Mars uniting with the earth so that iron was created. A particularly interesting example is zinc, based on Mars and Jupiter joining forces, so that 'live' tin was permeated with 'living' iron. This was at a stage in Earth evolution when metals were not yet solid but indeed still 'alive'.

The whole earth is dependent on events in the worlds of elemental spirits, the elements and the planets. You have to gain living insight into what happened when Mercury took over from Mars after the Mystery of Golgotha and new elements, new ideas and new inner attitudes came into existence. It will then be possible to be in touch with the world of nature, with the elemental spirits, the spirits of the seven free arts and the planetary system and get them to work with us in configuring something new, give birth to something new, let new developments arise.

Humanity had to ignore those spirits for a while in the course of world history, for it was necessary to concentrate wholly on the material world, which has increasingly been the case from the fifteenth century onwards. From about the middle of the twentieth century onwards it became apparent, increasingly so, that this is a dead end. We have to find a way out of it—or else perish.

Let me quote therefore from the notes K. Walther made when Rudolf Steiner visited the Guldesmühle on 12 August 1920.[4]

> The spirits of the elemental world—gnomes, undines, sylphs and salamanders—are actively involved in plant development. They used to be guided and influenced by higher spirits that have now withdrawn from this sphere of activity, just as they will at times withdraw their influence from human beings and turn to work at a higher level. The elemental spirits are then left to themselves and other spirits (Lucifer, Ahriman) take hold of them and distract them from their work in plant development. The result will be that the spiritual energies of plants decrease, gradually causing them to die off completely. Even artificial fertilizer will not help in this case.
>
> Efforts must be made to enter into the elemental world in a living way and seek to make a connection with these elemental spirits. We need to make friends again, as it were, with the elemental world and prevent other powers from using them. We must try and get these spirits to continue to serve plant growth.
>
> Farmers are able to develop powers of this kind, they will be priests in their work.
>
> People who do not succeed in making such a connection will find that yields and the quality of produce grow less and less within a few decades, and that there will be no help for this.
>
> Undines play a part in producing dew. Other spirits are about to distract them, too, from the work they have been doing until now. The result will be that no more dew is created as time goes on.
>
> Here, too, human beings will have to take a hand.
>
> Another thing human beings will have to do is to make a conscious relationship with the rhythm of the seasons. (Entering

into the life of nature with our physical body in spring, with our body of creative powers in summer, our sentient body in autumn and the I in winter.) If we develop this relationship, spring sowing will be done in a different mood than autumn sowing, for instance.

To gain a relationship in the spirit to the animal world, human beings must gain insight into the group souls of animal species. Animal instincts are fading even now, and will do so even more in the future. Thus animals will no longer leave poisonous plants in their fodder untouched, as until now, but eat them with the rest of their feed.

Gaining insight into the group souls, it will be possible, for instance, to balance out this loss of animal instincts; man can thus be a helper for the animals.

It is far from easy today to develop an inner attitude that allows one to be in touch with the elemental spirits and the group souls. This is why I am trying to show the ways Rudolf Steiner and Alanus ab Insulis wanted to open up for us, trying to help us gain insight into the spirits which are connected with the earth and love the earth.

It has to be accepted that it is not possible to bring a spiritual impulse to realization with the terminology used in modern science. We need to find other words, new ideas, different inner responses so that we may build a bridge for the elemental spirits, the group souls and the spirits of the hierarchies. We need to discover the powers that lie in the seasons. We need to relate to plants and animals in a way that is both loving and discerning, so that we may understand the way things happen in the etheric, in the astral world. And it certainly is difficult.

John the Baptist said: 'Seek a change of heart!'

This is what we need to do if we are indeed to find our way to the spiritual entities. People like Francis of Assisi, Empedocles, and even Faust were well on the way.

In her book *Erlöste Elemente*,[5] Johanna von Keyserlingk wrote of the tree spirits in the woods of an estate in Silesia imploring her to release them—nature spirits which Faust put under a spell in the Middle Ages.

These things we are told are real, they have happened and are one way of finding our way to the spiritual principle in the earth by developing an inner relationship to them.

When he visited the Guldesmühle, Rudolf Steiner spoke of the inevitable future the elemental spirits would have to face if human beings are unable to relate to them in a loving way.

Love is the element, the boat, that will take us to the shore where we can gain living experience of the earth and the spirits connected with it, if we take the road, the Easter Saturday road, to the centre of the earth which the ancients called Shamballa—'fairyland'. They would set out on this road to find the powers that would enable them to live on in a positive way in the world.

One day when Rudolf Steiner had talked about the group souls and the individual nature of a farm in the morning, he and my mother went for a walk along the broad avenue leading up to the manor house. My mother asked him: 'Is it really true, Dr Steiner, as it says in the Bible, that the disciples needed no more than a mustard grain of faith and they'd be able to move mountains?' Rudolf Steiner: 'Yes, that is really true.' My mother went on to ask: 'Is it really just like that, so that if one wanted to move Zopten hill, which we can see from here, one could indeed do so?' Rudolf Steiner: 'Yes, one can certainly do so. The faith we generate in our hearts can move such hills or mountains if one has the most intense wish to do so. Only you should not expect it to happen right away. It may perhaps be only in your next incarnation that the result will come, with the mountain then in another position.' People can use their powers of mind and spirit to intervene in nature and bring about changes, changes in the geography, for example, but they must awaken faith in their hearts. To breed something new, transforming a wild life form into a cultivar, is an act that needs this kind of faith if it is to succeed, and with it must go the greatest selflessness of which a human being is capable.

My mother would often tell me of that conversation, saying again and again how important it is to have this faith, developing it in our hearts. Even Peter did not have enough of it, but the Christ showed him the way.

Farmer Ernst Stegemann talked with Rudolf Steiner about breeding and food quality. Rudolf Meyer and Johanna von Keyserlingk were waiting for him afterwards, and Mr Stegemann was much moved as he told them about it. In his book *Wer war Rudolf Steiner?*,[6] Rudolf Meyer described the occasion as follows:

> I remember a day in Stuttgart when a friend came and told me, deeply moved, of a talk he had just been able to have with Rudolf Steiner. Running a large estate, he had been aware for decades that seed quality was progressively deteriorating. At the same time the food quality of crops also went down, in spite of yields increasing by the hundredweight. Mr Stegemann had taken these questions to our teacher. He received answers that were also to give him his life's mission. Rudolf Steiner had told him that all the cultivars providing our food would degenerate within a very short time. This was connected with the end of the Kali Yuga. New ones have to be bred. And he also gave him directions on how to breed a new cereal plant from particular grasses, a grain that could then be used to make a hearty bread.

Johanna von Keyserlingk made notes and later wrote:[7]

> Won't it be necessary for new powers coming from the world of the spirit to create new plant forms, with the aid of human beings?—Dr Steiner: yes. Or that existing plants are improved, among them also plants now considered weeds— Dr: yes.

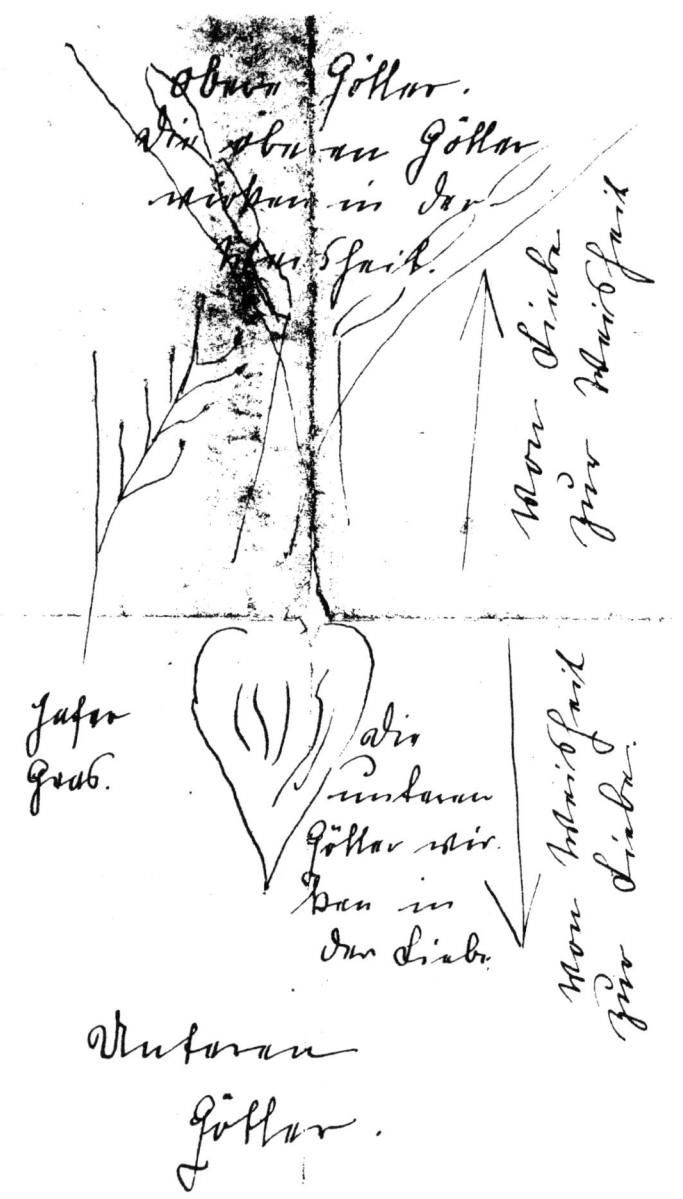

Notes made by Johanna v. Keyserlingk during a talk Rudolf Steiner had with Stegemann in Dornach in the autumn of 1924

Writing in the sketch (see left):

> upper gods ↑ from love to wisdom
>
> The upper gods
> work in wisdom

oat grass

> —The lower gods
> work in love
>
> ↓ from wisdom to love
> lower gods

'Let the sower carry love in his left hand as he does his work. In the right hand hope and faith.'[8]

It was also during this talk that Rudolf Steiner spoke of one-grained wheat (einkorn) and oat grass (presumably a *Bromus* species) as highly suitable for transformation.

Sasterhausen estate

People have often asked me what we did at Koberwitz after the course on agriculture.

I have to say that my father, Carl Count Keyserlingk, had to break away from the men who were leading the group for biodynamic experiments, young friends who wanted to rush ahead with enthusiasm but rather hastily, without first trying the new method out on different soils, before it was made public. He gave up his chairmanship of the group because he felt these friends were not really responsible in their aims.

In the years that followed the course, the firms IG-Farben and Kali-Syndikat had grown more and most hostile towards Carl Keyserlingk. Schisms developing in the family firm were past repair in the end, and finally he had so many problems that he left Koberwitz in 1928 to take over the Sasterhausen estate in Silesia. He died suddenly at the end of December

1928 when on his way to a conference in Dornach, one might say from a broken heart because of the way things were going.

My elder brother Wolfgang and I took over the Sasterhausen and Raben estates after our father's death. We divided up the management, for my brother was unable to accept the biodynamic approach to agriculture. I took responsibility for the forests, about 80 acres of park-like grassland and a few fields.

The pastures had not had artificial fertilizer on them and we were pleased to find that the preparations, which I made myself, and the compost proved extraordinarily effective.

Sasterhausen used to be the seat of the Cistercian Grüssau Monastery's abbot. It was large and spacious, with more than 40 rooms, a small chapel and a large room with a domed ceiling and walls 2 metres thick. The large park used for pasture extended to the south. There were fish ponds and ancient oaks, well over a hundred years old, as well as a sacrificial stone. The windows of my laboratory offered a great view of the park. Directly beneath the windows was the millstream that was the boundary for our experimental gardens. These were surrounded and protected by a high wall. We had 50 beehives in the experimental gardens. These would be taken out on the nearby heath in the summer. The bees were clearly happy there and gave us the most marvellous heather honey. I only got stung once in all my years there.

Next to my laboratory was a room with a fireplace where I did ashing experiments. Whole tree trunks could be burned there, in a vast chimney rising high above the roof, to study the flames. Every tree had its own kind of flame in both form and movement. Birds, mainly falcons and sparrow hawks, would sometimes fall down the chimney. I usually noticed and removed them from the chimney. My cousin Aki would look after the survivors.

It was a 'sunny delight' if we managed to keep these falcons alive and perhaps even train them to hunt, seeing them rise up into the sky and the light. We would then 'remember' earlier times—just as Rudolf Steiner had suggested in the lectures he'd given in Pforzheim.[9] We were able to 're-member' a monk doing breeding experiments in the small monastery garden to study the laws of heredity which later provided a basis for the work of Darwin. Recalling such things made us feel connected with the elemental spirits and the people, long since dead, who had been involved in the work in the past. And we had their help in our endeavours.

Joy was altogether an important element in our work, and I'd therefore like to say a few things about this basic mood, though I do find it extraordinarily difficult to put it into words.

We took delight in the falcons' flying and on Easter Sunday mornings in the fact that Mary, who to us was the 'earth mother', called the risen Christ a 'gardener'. She had stood by the cross with John on Good Friday. Medieval artists show John as the 'eagle' among the evangelists. We felt powers coming together here that can also stream into the free heart space which every human being has in his breast, 'where Lucifer and Ahriman cannot enter.'[10] We can use these powers to get in touch with the group souls of plants.

We were able to dry tons of medicinal plants, lots of Japanese peppermint, sage, and many others, in the manor's vast attics and deliver them to the Weleda. Vegetables were bottled and sold, and cartloads of cabbages sent to market.

When the war started I was given government permission to make jam with a combination of wild and garden fruits in an effort to use more wild fruits for food. One outstanding product was an elderberry jelly; we added the elder stalks to the mixture, as they contain a great deal of pectin, as well as lemons and plums.

We also found that bottled fruit kept better than tinned. If we took the positions of the moon into consideration, our fruit would keep very much better if bottled during the waxing rather than the waning moon phase. These were discoveries made more or less by the way in the vast monastery kitchen.

I was employing three women to help. We were such a good team that old Babuka and Gustel Krämer and his wife came with me to Wahlwies when we started the first German children's village in 1946.

In the large field behind the wall grew raspberry, currant and gooseberry bushes, with medicinal plants between the rows. We also planted potatoes there which I had brought with me from Ecuador. These were always specially prepared, not planting whole potatoes but only eyes cut out from the tubers in spring and planted. We found from these experiments that the potatoes would not deteriorate if treated in this way but get younger and fresher.

Medicinal plants, above all peppermint and sage, had a powerful aromatizing effect on all plants growing around them. Symbiosis with horse-radish also had a positive effect. We planted it not only at the four corners, as suggested by Rudolf Steiner, but all around the field and also in the rows themselves. It absolutely vitalized those potatoes. But all cabbage varieties also showed a highly positive reaction to horse-radish. We would always create an enclosed 'space' where horse-radish would be active, or peppermint, or also raspberries. Such spaces would be created all around the plants intended for breeding.

We had started to plant maize in the form of hedges. Maize is the Saturnine member of the *Gramineae*. Planted all around a field it can be highly effective, not only as a windshield but also as a Saturnine influence on its 'inner space'. Saturnine powers are evident in the unusual feature in which the plant holds its stamens up high, offering them to the heavens in a wonderful sea of flowers, whilst the pistils, drawn to the earth, develop low down on the stem. A space is created in the horizontal as stamens and pistil are held apart in a gesture that can also be seen in the ringed planet.

We tried to expose specific plants to planetary influences by finding old tree trunks where the heartwood had turned to earth but the bark and outer wood were still live, cutting them into handy sections and putting these in our experimental gardens. We sowed cereal grains in the rotted-down soil in these 'baskets'. Taking the stump of a plum tree and sowing wheat in it, for example, exposes the wheat to a Saturnine principle. (Plum is a Saturn tree.) Both seed and fruit showed distinct changes, with trees relating to different planets always showing different effects.

I gained the impression that the perennial mountain rye I had been sent by friends in the Urals responded particularly well to Saturnine influences, which were evident throughout the vegetative period and particularly also at harvest. I noted very powerful influences when sowing in oak stumps.

An old willow trunk I had brought had grown almost horizontally. We were able to halve it lengthways, filled the trough with moist decayed material and kept this well watered. I sowed mountain rice in it which developed in a very peculiar way.

Yet before we were able to record real results, war came to the area with all it entailed. The soldiers used all our photographs, all records, and indeed any paper they found to light their fires.

When I had been working for a year with Ehrenfried

Pfeiffer, Mr Eckstein and Miss Riese, people had become aware of my interest in these issues. Many friends then brought wild plants or local varieties from their travels for me to work on. Dr Wegman and Zeylmans van Emmichoven brought local varieties from Sinai, for instance. These did very well and I still have an ear. Pfeiffer gave me his 'booty' from Rome, a one-grained wheat which still grew wild in the ancient ruins there. He had brought me some handsome specimens for acclimatization and development. We worked with this plant for many years, facing many problems and difficulties, but to no great avail. Pfeiffer himself did later manage to get a transformation.

Dr Wegman brought me some wheat from the Burgenland area in Austria, a place where the Hibernian Mysteries were thought to have been and where the Templar castle Bernstein rises above old antimony ore mines. When she came to visit us in Sasterhausen in 1930, before the devastating annual general meeting at which she was forced to withdraw from the Dornach Council, she asked many questions about the seed she had sent and about our experimental work, especially the work done on the Chinese sweet potato (*Dioscorea batatas*), which Rudolf Steiner had suggested should be acclimatized.

She took a tremendous interest in our work and told us that it had been Rudolf Steiner's intention to make agriculture and the Koberwitz impulse part of the Medical Section rather than the Science Section, for he said it was the Medical Section which had to take care of the work to heal the earth. In his view, the Koberwitz impulse was an effort to revitalize the earth and not so much a scientific issue.

When working for my diploma in agriculture, I took part in a field trip organized by the breeding institute at Breslau University during which our professor discovered a four-rowed couch-grass. He was just as thrilled as I was! I dug up some specimens and planted this sensational plant in my gardens. I did not manage to change it, however, before the end of the war came and we had to leave. It is possible that it might still be found at Sasterhausen—between the rose garden by the terrace and the wall on the farm side.

A meditation which Rudolf Steiner had given to Mr Stegemann became a very real part of our work in those years. We would enter wholly into it, with a small pile of a few hundred grains before us, finally putting perhaps just a single one of them into the soil. This would be the grain which had given a 'response', giving the impression that its aura had changed. Elsewhere Rudolf Steiner also spoke of 'small blue flames' that may rise above a particular seed grain when one is working with it meditatively.

Learning to work in this way we create places where people who are between death and rebirth can bring their powers into play and take our work further. This leads to transformations in the plant world which then need further care to 'stabilize' them, making them into reliable cultivars we can offer to farmers.

It was in those years that 12 different forms were created from a single seed in Hugo Erbe's establishment and also Immanuel Voegele's place. Voegele, for example, produced 12 different phenotypes from a spelt [German wheat, *dinkel*]: awned, with spadix, with a thick head, etc. Professor Bier in Berlin heard of this. He had achieved remarkable things in his Sauen Wood in the Brandenburg province.[11,12] The professor visited Voegele in Pilgramshain to look at his mutations. It is reported that he had clear-felled pine woods of long standing and then had the ground ploughed up. Masses of yellow lupin seeds came to the surface and germinated, against all expectations, but then developed into a number of different lupin varieties.

Differences between East and West in the earth organism

When Wolfgang Wachsmuth returned from an extended trip to China in the early 1920s he discussed this with Rudolf Steiner. Apart from many other things, Rudolf Steiner told him it was important to acclimatize and develop the Chinese sweet potato in Central Europe, for strong light powers could be taken in when it was used for food. In the course of this conversation, Rudolf Steiner drew a curve that started high up in the East, moved horizontally across Europe and then turned downward as it reached America in the West. He said these were the light powers (not the sun's rays); they came in more strongly from the cosmos in the East, where plants would take them up. In Central Europe these powers moved in a direction horizontal to the ground, and in the West they went below ground, down into the depths, becoming electricity. People would, however, need food plants that could take in more light, and so it would be necessary to acclimatize these plants in Europe.

Rudolf Steiner's suggestions have been in our hearts and minds since 1925. Ehrenfried Pfeiffer would return to them again and again. Mrs Wund and Mrs von Grone had all kinds of potatoes sent from China, a big box full. But once planted they would always turn out to be varieties that had been imported to China from the West, from Britain and America.

The sweet potato used as a food plant in southern Europe today, *Ipomoea batatas*, has also been introduced from the West. It looks like a long, large potato, has a high water content and will freeze at as little as 1 or 2°C. The boiled tubers have a sweetish, floury taste. With their high water content they are not able to take in light powers, and so this cannot be 'the potato' Rudolf Steiner mentioned. The others were Solanaceae, members of the deadly nightshade family, the very opposite of vehicles for light.

A great deal of effort finally led to a yam native to China's high plateaux, *Dioscorea batatas*, sweet potato. I found it by chance in the Hohenheim botanical gardens and was finally able to get hold of the plant in Dahlem.

Dioscorea batatas is a climber with a marked spiralling tendency, unable to grow erect on its own. It needs supports on which to grow, with the top end hanging down loosely. Given this protection, it produces fine white flowers in spikes. These have a vanilla-like scent and attract many insects and butterflies. The seeds look rather like beechnuts—small, brown grains in a trimeric involucre. Unfortu-

Dioscorea batatas, *family IV*
700/23

nately we have not so far succeeded in getting them to ripen and be capable of germination in our part of the world. The small round brown nodules that develop in the leaf axils do, however, germinate easily in a hotbed, making vegetative reproduction possible. This also works very well with small pieces of the root.

This plant, also called sweet potato, produces thick roots, 5–20 cm in diameter and more than 1.5 metres in length. The white inner part is surrounded by a thin, pale brown skin with a very few fibrous rootlets. It keeps well through the winter if stored in the soil or in a cellar. It dries out in warm air and can then be used to make flour. Boiled it is a crumbly, dry mass, slightly sweet and tasting pleasantly nutty, which combines well with almost all other foods. Compared to the American sweet potato, this root is not frost sensitive.

We had to ask ourselves how people in China harvest roots that go down so deep, but found that the soil layer in the

Dioscorea batatas, *family IV*

plant's native habitat is so thin that the roots are forced to grow more or less horizontally. Where it would be possible for them to go down deep, the Chinese put large pieces of slate underneath to prevent vertical growth.

Breeding experiments, first at my place in Silesia and later at Miss Windeck's by lake Ammersee, always showed the same behaviour. Large numbers of nodules developed in the leaf axils and would germinate very well in a good light and a hotbed. They could be planted out in their second year. Depending on the support provided, they grow 1 or 2 metres wide and 2 or 3 metres high, using every opportunity to wind in a spiral and widen out up above, whilst the thick root goes down deep.

It would of course not be cost-effective to have to dig them

Dioscorea batatas, *family IV* 700/34

out laboriously. All kinds of things were tried to make them grow horizontally. But they would split wooden boards which would rot away and grow soft, so that the roots grew through them. The pieces of slate I was able to obtain were too small. The roots would again and again find a gap; they'd even grow thin and flat to pass through and then thicken up again below. We did only rarely manage to overcome this need to go down. Generative reproduction was not possible, as we could not get the seeds to ripen. In the two or three years when we did successfully grow the plants, we did not manage to change the tap root and make it shorter and thicker by using quartz meal or the Horn Silica preparation (501). Better results were seen with planting out in, and surrounding the seedlings with, pure Cow Manure preparation (500). At that point, however, the war put an abrupt end to our work. I went to the front, and soldiers' boots trampled my experimental gardens into the ground. It seemed that the work done with such love for many years had yielded nothing at all.

Will it ever be possible to get those 'cosmic' roots to assume a spherical form? I'd like to come back to this later.

In the talk he had with Mr Wachsmuth, Rudolf Steiner suggested acclimatizing the plant in Europe because in a special way it was a strong vehicle for light. If we think about the possible meaning of the line Rudolf Steiner drew on that occasion, new prospects open up that may lead to completely new insights. The line was meant to show that in the East streams of light come down to the earth from up high in the cosmos; in the middle, that is in Europe, they are more or less horizontal to the earth's surface, and in the West, in America, they go underground where they turn into electricity. Study of Rudolf Steiner's agriculture course makes it clear that the way the light streams go in Europe also corresponds to the 'diaphragm' of the earth (solar plexus).

To come closer to our theme, let us consider the following. The suggestion made in the agriculture course is that farmers should make their farms into organisms in which a 'farm individuality' might incarnate the way an I incarnates in a human organism. It is, however, fundamental to realize that such a farm organism differs from a human being in that its 'head' faces down, its 'metabolism' is raised aloft. We only need to look at a plant to see how the roots grow below the earth's surface whilst the reproductive organs are raised up to be open to the cosmos. We may thus say that all human and animal life on a farm takes place in its 'metabolic organization', whilst the plant roots, which are comparable to the organization of head and senses, are active below ground. Like a diaphragm, the soil is a border that separates them, with cosmic and terrestrial forces passing through it as they work together and are kept in circulation

The whole farm organism thus lives in a way that is the polar opposite of man, standing on its head, as it were. We might also say that the human being is an upside-down plant.

This realization has had major consequences, above all in medicine. To understand the significance of the diaphragm in this context even better, let me try and describe it.

It is like a bell that is ringing for the inner human being. The muscles of it are fixed below but move freely under the xiphoid cartilage of the sternum. The muscular wall has three main apertures through which the upper 'head' part of the human being connects with his lower 'metabolic' part. As already stated, it is the other way round for a farm organism.

One is the oesophageal aperture. Food taken in through the mouth goes down through the oesophagus into the stomach and to the abdominal glands. Another opening is for the vena cava and the third for the aorta. The blood circulating in those vessels helps to keep human beings strong and mobile. The apex of the heart is mirrored by the liver and gall-bladder on the other side. The diaphragm can thus be seen as an ideal bell in the human being. An inner circulation develops in which sensitive individuals may with the help of the solar plexus perceive processes that occur within the earth and let them become living experience. With practice, he can thus influence his breathing and metabolism. The solar plexus plays a very important role in sleep and during meditation.

The significance of the 'earth diaphragm' shows itself when we find animal species that have a particularly sinewy, tough diaphragm—cats, for example. Other animals, like almost all birds, do not have one at all. Knowing such things helps the farmer to see his farm as a harmonious organism. In the 'earth head', roots are actively exchanging salts; in the 'metabolism' above ground, stems, leaves and flowers develop to exchange oxygen and carbon dioxide, heat and light.

The fertile soil in the fields is the 'rhythmic system', the breathing, where cosmic and terrestrial forces circulate. Here too, the diaphragm, which is the soil surface, separates head and root processes from metabolic processes.

The line indicating light qualities in different parts of the world runs in the same direction as the earth diaphragm in Europe. It is different in the East and the West. We can sense that 'East' and 'West' are not merely geographical terms here but—as can be seen from the terms Orient and Occident—refer to qualities. This is most beautifully expressed in the *Foundation Stone Meditation*.

The West truly ends at the mountain ranges that run from north to south in America, whilst the East has its limit in the Pacific Ocean's vast water masses. In this space, where no continents are, etheric streams and forces can act without hindrance between earth and cosmos. Going through the Panama Canal, one really feels the difference between the

Gulf of Mexico and that wide, peaceful ocean. The difference is also apparent in the different cultures there. In the wide expanses of the Pacific, the traveller is aware of the aura which is open to the spirit. The peoples living to the west of the Cordilleras have a completely different potential from those who live in North and South America.

The contrast between the form-giving powers in the West and the metabolic forces in the East is also apparent in mythology and religion, especially if one considers the animals that were venerated. In the West, priests and chiefs wore feathered head-dresses; in the East, the dragon, an earth creature, was important.

Other contrasts may also be seen. Giant trees with the mightiest trunks on earth grow in America, whereas the cherry blossom is the glory of Japan. Maize, tomato and potato have come from America; rice, sugar-cane and our *Dioscorea batatas* from Asia. The children and grandchildren of Europeans who have emigrated to America are all taller than their forebears, and anyone asked to draw an American and an Asian would no doubt show a blond giant side by side with a dark, finely made Asian. The habit Americans have of putting their feet on the table is not simply bad manners. It merely confirms what has been said about the way light acts in the West. As soon as light enters into terrestrial regions it becomes electricity. This can produce sensations in the feet which people want to avoid. An Asian, on the other hand, likes to sit on the ground and will kiss the soil if he feels specially connected with it.

People also think in a different way in East and West. Americans will immediately relate their thoughts to outer reality. They have thought up and built the biggest bridge in the world, the tallest skyscraper—these are form-giving powers related to the earth. In the East, on the other hand, souls feel the fire of enthusiasm for ideals, for religion, for spiritual goals. For centuries, the people in India venerated the most sublime wisdom so much that they never got to the point of bringing it to realization in the physical world.

All this has to do with the angle taken by the powers of light and with the earth's diaphragm. In the West, plants must first with much effort grow up into the metabolic sphere from the zone of head and root processes that lies below. Giant trees result. Other plants, such as tomato, stay in the earth-bound zone, or they suck the leaf process down into the soil, like the potato in producing its tubers.

Maize also goes down into earth-type processes. Only parts of the flower, the stamens, rise up into the flowering region; the other parts remain in the sphere of terrestrial forces, with the ovary and pistil drawn down into them, with seed development low down on the stem. As part of a staple diet, the heavy, Saturnine cobs have a hardening effect, mainly on the thinking.

In the East, roots must also develop in the ground, but the plants need to make little effort in reaching for the sun; they soon reach the 'light region'. In East Asia, plants therefore develop above all in the 'sap region', with sugar produced in leaves and stems. Aromatic teas and sugar-cane come from those parts, and rice actually grows in water without rotting, for their flowering processes are extremely powerful. Tea, sugar and rice will therefore strengthen the 'flowering processes' in humans.

These few observations make us immediately aware of the contrasts created on earth.

In the middle zone, in Europe, the polar opposites are in balance. On the other side of the globe, in the Pacific area, where there are no continents and no contrasts, the earth's surface lies far below the surface of the ocean, closer to the centre of the earth. Above it lies the tremendous fluid layer where Mercurial processes mediate between hardening root processes (Sal) and flowering processes (Sulphur). The area is so vast, calm and open that the etheric principle can come into force. As at the border created by the human diaphragm, processes of etheric life move to and fro between earth and cosmos. A 'breathing' process arises in which the hierarchies are able to ascend and descend in a zone that has been calm and peaceful from the very beginning, for the catastrophes that put an end to fiery Lemuria in the East and later to form-giving Atlantis in the West did not touch the calm region of the Pacific Ocean.

At the centre of earth evolution, however, in a point where everything comes together, we have—in terms of time—the Golgotha event, the turning-point of time. The Christ Sun had united with the sun, creating a spiritual focus at its centre from which a new future can develop. The Christ's Sun powers, which had radiated from the cosmos down to the earth from time immemorial, have since been acting from this centre. This made the whole earth into the body of Christ, with his ether body now working in the earth's fluid organism. Rudolf Steiner spoke of this in a lecture called *The Etherisation of the Blood*.[13]

Up to that turning-point in time Sun powers streamed down to earth irrespective of whether people received them consciously as givers of life. Rudolf Steiner tried to show that now, when the Mystery of Golgotha has happened, the Christ powers no longer stream down from the sun but flow out into the cosmos from the inner earth. Human beings can have a part in this if they let the Christ come alive in their conscious minds. If we give thought to the Christ reality during our waking hours, his powers will enter into us during the night and give us new life. Processes from the daytime waking hours create the opening that comes during the night.

The same holds true for every farm-self, the individual nature of a farm, if the people who are part of it take the

Christ and Michael, 'his countenance', into their hearts and minds.

A farmer who knows nothing of etheric activities or of cosmic powers and because of this goes against them in his work is on the other hand sinning against the future of this earth.

These powers of life come to the earth organism as a whole in the Pacific region, an area where there are no continents and an exchange can take place between earth and cosmos. From there comes the Christ Sun we read of in the Bible. It comes through Michael and is like a second Christ event. It originates in the north-west, and if we gain insight into these things we come to see another line which is like a mirror image and complement to the above-mentioned light line. Let us call it the Michael line. Its etheric activity thus arises above the Pacific Ocean and comes down towards the earth all the way to the East. Christ-filled powers of insight enter into the darkness with it, and it is connected with the Christ's coming in the ether which 'will be like a natural event'.

The Bible speaks of this as follows (Matthew 24:30): 'And then shall appear the sign of the Son of man in heaven: and then shall all the tribes of the earth mourn, and they shall see the Son of man coming in the clouds of heaven with power and great glory.' And in Luke (21:27) we find: 'And then shall they see the Son of man coming in a cloud with power and great glory.'[14]

Rudolf Steiner spoke of the same event, saying that the whole of world evolution will depend on whether it will be possible to take the light of the West to the East. For the East, now in danger of drowning in materialism, needs this ether light, these Christ-filled powers of human insight, for only they will take it on into the future. The most ancient spirituality of the East needs thoughts and insights that will rejuvenate and revitalize it so that present-day materialism shall be overcome.

Thoughts like these are not intended to increase our knowledge but to strengthen our hearts; for growing and breeding plants also means to gain insight into the essential nature of the earth, its processes and laws, and to take them seriously.

Biodynamic preparations

Stag's bladder and yarrow

This preparation was also used in the plant breeding work done at Sasterhausen.

The gesture we see in a stag's antlers is one of opening up towards the heavens. But he is actually doing this with his bloodstream, for his antlers do not develop from skin that hardens, as in the cow, but out of the bloodstream. A cow's horns and claws regulate the vital processes in the inner organism, whilst the stag is able, having antlers, to take the surrounding world, including the influences that come from the outer cosmos, into himself.

The stag takes these influences into itself in an A-gesture, opening up to the world; this then becomes an E-gesture in the kidney, and in the bladder finally an O-gesture. The process is open to an artistic approach. There is a flowing movement from antler to bladder and the latter can thus be the protective vessel in which yarrow responds to influences in a way where 'the spirit always wets its finger with sulphur when it wants to take the different substances—carbon, nitrogen, etc.—to their appropriate places in the organism'. (Rudolf Steiner, *Agriculture*.)

Yarrow is therefore also very helpful planted around cereal crops or vegetable beds to create a special 'space'.

Following the whole process from A to E and O inwardly, the question may arise if this would not also have an effect on seed. We have filled a bladder with fodder beet seed and let it go through the same process as yarrow, from St John's-tide to Michaelmas and through the winter until Easter, above and below ground. We then sowed the seed. The beet grew so

*Fodder beet harvest on Sasterhausen estate in Silesia in 1937
Beet seed pre-treated in stag's bladder*

large that one needed two hands to lift one of them from the soil. The result was quite evident and could have been developed further, offering major economic advantages.

Stinging nettle

This plant regulates iron activities in the soil. It is important to grow it for 'it makes the soil rational', meaning that it enables plants growing in it to draw on the whole environment for the powers they need. Taken to a higher level by making them into a preparation, nettles are a substance that makes the soil 'individualize itself' for the plants one wants to grow in it.

Butterfly caterpillars and also pupae are often found on the underside of nettle leaves. Once they are butterflies they can then take the 'bright, rational' nettle impulses into the environment. This plant, too, can be used to create 'spaces' that give health, for instance where crops are affected by smut. A strip of nettles between barberry bushes and an affected crop gave excellent results.

We can study the nettle actions on our own bodies. Used therapeutically by applying it to the skin it will cause slightly painful rhythmical movements in the tissues which persist after removing the compress. Excessive activity or spasms in the head (migraine) can thus be drawn off into other parts of the body by using nettle, thus restoring harmony. Parts of the body that have hardened and are dying off can be revitalized by this plant.

When a blacksmith hammers red-hot iron, the fiery metal on his anvil sends out sparks. In a similar way, enlivening movement is generated when nettle comes in contact with the human skin or also the soil. (Iron hammer scale is an important medicament in anthroposophical medicine.) Organ processes that are in the process of dropping out of the life sphere are reintegrated in such a way that rhythm develops in the living body, sometimes so powerfully that the heart rhythm will change, and this may persist for a long time.

Nettle thus generates movement and conscious awareness wherever you put it. Breeders who understand its essential nature therefore find it a most important plant.

Peritoneum and dandelion

Rudolf Steiner would often visit the upper classes at the Waldorf School, a great pleasure for teachers and pupils, for he would usually talk about important issues in life. On one occasion he came over from the Institute of Biology which was next to the school and told of his work with Dr Kolisko. We knew her well, for we had been allowed to do capillary dynamolysis in her laboratory. She had had a pit dug 6 metres deep in the garden next to her workrooms and there she let crystals grow so that she could study the influences of the moon and the planets.[15] One day she almost fainted down in the pit and did not have the strength to get out of it. Having missed seeing her, we came and were able to get her out just in time.

Rudolf Steiner told our class that it had finally proved possible to demonstrate the existence of the etheric world experimentally and that even materialistic scientific thinkers could no longer deny this, for the experiments could always be repeated. I have never seen Rudolf Steiner as pleased as he was on the day when he was able to tell us of the success achieved with capillary dynamolysis. I remember him saying, in response to a question, that the handsome sunflowers growing at the Institute should really be called Jupiter flowers, for they belonged to Jupiter, which proved to be true when Dr Kolisko 'fertilized' them with different tin potencies.

When he gave his course on agriculture, Rudolf Steiner said that Jupiter was active in dandelion when it made the grasslands golden in spring with its flowers. Yellow, he said, was Jupiter's colour, and when dandelion developed its seeds, a whole round world came into being. Children liked to blow this away, but it really showed the O-gesture we knew in eurythmy, which was also a Jupiter movement. A movement that imitates the universe and receives it into itself.

This led us to work with Jupiter and tin in the plant world. The O-gesture revealed the power of Jupiter, seeking to shape the surrounding world out of wisdom.

Dr Eugen Kolisko, my teacher, had given a course for pupils who wanted to study medicine. For a whole week one of the seven metals and its actions were discussed, a different one on each day. He would always say: 'Take note of how the five preparations used in agriculture correspond to the five

Experimental gardens of the Institute of Biology at the Goetheanum

endocrine glands in man. If you put them into your compost heap, this will make it into an organism.'

One day he brought a cigar box into class with spheres of mercury rolling around in it. He sprinkled some finely powdered tin over these. The Jupiter metal immediately formed a skin around the Mercury metal, but as a 'tube', with the small mercury spheres moving up and down in it. This gave us the image of the intestinal system. He said: 'This is the function of Jupiter in conjunction with Mercury. And if you have a patient with ileus, a blocked intestine, you give him an injection of tin in high potency (20–30x) and also mercury (5x). This will resolve the blockage and heal the condition.' I have often used this gentle treatment later on, even in a case of ileus caused by eelworm.

I recently heard that someone making the preparations did not know if he should take a diaphragm or a peritoneum for the dandelion preparation, thinking it probably did not matter and he might use either. The two are utterly different, however. The diaphragm separates the heart and lung region and the neurosensory system from the metabolic region. The peritoneum on the other hand envelops the latter, that is, the intestines. The functions of peritoneum and diaphragm are thus quite the opposite.

The bovine mesentery is the part of the peritoneum from which the intestines are suspended in the abdominal cavity. The peritoneum is a thin membrane lining the cavity; it has grown together at the top, hanging down into the cavity. This latter part of it is reticulate, with large lymph vessels passing through it, which gives it the reticulate or netlike appearance. This part of the peritoneum is used to envelop the dandelion preparation.

When I was at a field dressing station during the war, a soldier came to me one night who had a terrible pain in his belly. I put him to bed with hot and cold compresses and diagnosed a perforated gastric or duodenal ulcer. Knowing about the peritoneum's enveloping function, I immobilized him completely. After two days the pain had gone. The peritoneum had protectively covered the perforation. The same happened with a patient who did not want to have an operation. There, too, the peritoneum caused the problem to heal itself, closing off the ulcer, so that the inflammation could subside. It is however important to see that the patient gets hardly anything to drink so that the food stays a firm paste and cannot run through into the abdominal cavity.

Dandelion, a plant whose seeds grow in a perfect sphere, has flowers as golden as the sun. Rudolf Steiner once called it 'a true gift of heaven'. Sown up in mesentery, wilted dandelion surrounds itself with an earthly body; it is born in the soil, as it were.

We can experience this progression from gift of heaven to child of earth in eurythmy by doing the I, O and A. The I coming down from heaven, the O enveloping the earthly element, and the A radiating out into the surrounding world.

Perhaps we might also think of the tale of the Star Dollar Child gathering the stars that fall from heaven in its shift when they have become gold coins.

Ashing

The large fireplace in our Sasterhausen laboratory was excellent for making the ash preparations. I just had to make sure sufficient seeds and skins were to hand. We used different sieves to get the charlock seed from the thrashing machines clean, removing all wheat grains. The seeds were then ashed. We did small-scale experiments first in experimental beds and later used the method in the field, with the help of Miss Hasché (later in charge of the observatory in Dornach). A whole bucketful of charlock seed was burned for one field experiment. Before that, we had already established that the seeds must not remain in an incandescent state for long in the ashing process. The temperature had to be kept relatively low so that the incandescent phase was missed out and a browny ash would remain. Years earlier I had used the process on a chicken farm in England and found that the power which is the opposite of the germinating power is lost at red heat.

The 6-ha field looked after by Miss Hasché soon showed results. A line as straight as if drawn with a ruler went down the middle of it. One half, where the 'pepper' had been scattered, was completely free of charlock, the other half was soon full of the bright yellow flowers. Sadly, the photographs taken at the time were lost during the war.

The experiments with mice, such a plague on many farms, were most interesting. I was lucky to have a lady on the staff who did not mind skinning mice and rats and putting them on stretchers to dry. We had many mice in our fields and animal houses. I offered the children a penny for every mouse and soon had not only sufficient skins but also the cadavers. These I put into a barrel to rot down completely. This 'compost' was then brought out together with the ash from the skins. Soon not a single mouse was to be found anywhere in the manor house or the experimental gardens, which covered quite a large area.

For another experiment I had concrete boxes made with subdivisions. The top and front of each compartment was open and we put in wire mesh, giving us a row of cages separated by concrete partitions that were about two fingers in thickness. For the experiment, a bag filled with mouse pepper was suspended from the top of the first cage on the left. The mice put in the cages—there were up to 10 of them—were of course all given the same food. We wanted to know if and how the suspended pepper would have an effect on them. The animals in the cages soon died in a sequence that went from left to right. Experiments with the ash of white mice gave a similar result, though not quite as definite. We also scattered the ash of rat skins and soon no longer had any rats in the gardens, nor any water voles, though the mill stream had been full of these for a long time. Rats continued as before, however, in the poultry houses that were about 1½ kilometres away.

It is difficult to get distinct results in open territory. The experiments I did at that time did however show that:

1) ash taken through the red-hot stage has no effect, whilst ash that is still slightly brown does, i.e. it still contains carbon;
2) the distance between ash and creature plays a role;
3) rotted-down cadavers also have an effect.

Getting to know the pioneer plant breeders

In 1926 my father sent me to Davos for a year for my health. I was supposed to prepare for my school-leaving examinations at a German school there. When I found out that my mother had written to the headmaster that I might actually stay for more than a year, I simply packed my bags and went to Dornach to see Dr Wegman. I passed the examination some years later in Breslau.

This decision led to my meeting Ita Wegman and scientists like Rudolf Hauschka and Mr Kaphan who was working with him. These meetings were of tremendous importance and would never have been possible if I had left it till later. I worked with Eckstein and Ehrenfried Pfeiffer in the laboratory under the Glass House, where I was involved in the early beginnings of crystallization experiments with sodium sulphate and copper chloride.[16] At the Dornach training school for young people, led by Pfeiffer and Maria Röschl at the time, I learned about all the research work done in the 1920s.

What I remember most of my talks with Dr Wegman is the intensity with which she spoke of her collaboration with Dr Steiner. To my rather naive question how they had handled the writing of the book *Extending Practical Medicine*, she replied: 'Sometimes he'd say a sentence, and then again I would.' That is how the manuscript was produced at the small desk in the Studio.

She was always able to see things in the widest possible context, above all how important it was to relate to the evolution of the earth, giving it new life and maintaining it, which would in future depend on human beings only. She cared as deeply about the further development of the

Dr Ita Wegman
1876–1943

Koberwitz impulse as the farmers did. In many talks with her and the people who were working with her I really came to understand how, as anthroposophists, we must include spiritual entities, the elemental spirits, group souls and planetary powers in everything we do and plan to do in this world.

Another question that was considered was that of having a rite for the medical profession, something that seems to have been completely forgotten today. The esoteric lesson given in Koberwitz on Whit Sunday had come about in such a way because of the questions the young people put to Rudolf Steiner.

Now that I am an old man I can openly say that Ita Wegman was so deeply connected with the esoteric impulses for our breeding work that she will always be there to help us if we turn to her.

The approach to work was very different with Ehrenfried Pfeiffer. He was doing basic research in the laboratory under the Glass House. With his crystallization experiments, first with sodium sulphate and later with copper chloride, we had to try and gain insight into the essential nature and the powers of light and of weather conditions. The results obtained with crystallization were less rich and varied than those one got with capillary dynamolysis, but they were very much more accurate with reference to specific issues.

Pfeiffer and Erika Riese ran plant breeding experiments in the laboratory garden, above all with one-grained wheat (einkorn, *Triticum monococcum*), oat grass (a *Bromus* species) and wild barley. In talks at which Mr and Mrs von Grone were also present I came to realize that we would only achieve something if we also took the spiritual entities into account. Pfeiffer would say over and over again that we must train ourselves to gain a real relationship to them.

Pfeiffer later worked on the processing and utilization of waste matter from large cities in the USA. His method was used in Nuremberg in Germany in the 1950s.

His friend Mr Schiller was experimenting in the Boiler House laboratory with recitation to influence flow movements in liquids, air and flames. We were able to see flames assume very different shapes when different people were reciting.

It was when he was still in Dornach, before he went to the USA, that Ehrenfried Pfeiffer succeeded in setting Roman einkorn in motion to such effect that Mrs Künzel

Biodynamic agriculture conference 1932, in Bad Saarow. At the back, from the left: Hans-Jürgen Senfft from Pilsach, Franz Dreidax, Gust Stegemann, Erhard Bartsch, Carl Grund, Ehrenfried Pfeiffer, Benno von Heiniz, Günther Wachsmuth, Walter Birkizt; front row, from the left: Moritz Bartsch, Josef Werr, Immanuel Voegele, Helmuth Bartsch, Rudolf Koch, Max Karl Schwarz

and Miss Windeck were later able to continue the work successfully.

Immanuel Voegele, manager of the Der Kommende Tag farming estates at the time, was also in Dornach. Rudolf Steiner, he and Pfeiffer had made and buried the preparations when the first experiments were done before the course at Koberwitz. I remember Rudolf Steiner getting really annoyed when a long search was needed before we found the preparations in the spring that had been buried there the previous autumn.

Voegele later took over the farming at Pilgramshain in Silesia where Karl König had established the first anthroposophical special education institute. I attended a medical course there, during which I saw Voegele's plant breeding work. He had had his first successful results there, working with Miss Windeck. Mr Voegele had a stiff knee from a war wound and therefore had to ride a horse if he was to get everywhere he was supposed to be, and I really envied him for this at the time. He later took on the management of the large estates at Schloss Hamborn where Mr Pickert, who had formerly worked with Ita Wegman, Mr Goyert and others, established the well-known residential school.

It was a time of almost breathless progress, with anthroposophical initiatives gaining far-reaching influence.

The first miracle, a gift from the group soul of cereal plants, came in Voegele's experimental gardens in Pilgramshain. Twelve different ears developed from a single grain, each showing a different form. I know that Mr Voegele would meditate on his plants.

I got to know Professor August Bier in Pilgramshain. He had cleared a dry area of its 50-year old pines. When the area was ploughed up, lupin seed that had lain dormant for those 50 years came to the surface and germinated. Only *Lupinus officinalis* had been sown originally, but the ploughed-up seed produced all the lupin varieties. Bier, who had sent a report to the Botanical Institute in Dahlem, spoke of 'delayed germination'. It will be for the reader to look for an explanation.

During a visit to his 'Sauen Woods', Bier showed us an area where he had left brushwood and needles lying as ground cover for years. He had dammed up streams and small rivulets and jays had soon sown numerous acorns there. In just a few years the dry pine woods had been transformed into a healthy woodland with a mixture of hardwood trees. Voegele had told August Bier of his 'wheat miracle' and the professor came to see it for himself within a few days. I was a

Immanuel Voegele

trainee at Pilgramshain then, and want to say that all of us—humans, animals and, I truly believe, also the plants—had tremendous trust in Immanuel Voegele. Our loving nickname for him was Manu.

With Hugo Erbe I came to know a completely different way of working with the plants and the elements. He used the power of his singing voice! Once when I visited him he showed me how he was able to shatter a glass with his voice. Another time, when it was very cold outside, he made us breathe on the window pane in a warm room. Ice flowers formed, but in different forms for each of us. Erbe wiped the pane clear and we breathed on it once more. Again each of us had his own kind of ice flowers. We realized that apart from its waves, moisture content and the like, our breath, and also our speech and singing, held quite different powers.

Erbe connected with the group souls in his singing, but he also found other ways of influencing plants, for instance with his special preparations. I'd like to give just one example here—a preparation made from red grape juice and home-baked bread. He would stir it and use it to spray plants and also dress seeds with it. Basing himself on the preparations Rudolf Steiner had spoken of at the Koberwitz course he developed a whole series of 'food and drink for the elemental spirit work force', as he called it.

Erbe would meditate, sing and pray, and after many experiments produced a variety of *Triticum caninum* similar to spelt in the shape of the ear. A number of different wheats came from this in the years that followed, including his 'golden corn'.

This was on his small farm in Roggenbeuren on Lake Constance. Later, I saw a large bed at his Thalhamer farm in Bavaria where he had sown many different grain plants, some of them given to him by friends, others he had obtained in his own experiments. He would take different plants from the bed every year and work on them. Before sunrise every day he would spray the bed with one of his preparations. Erbe's first successful transformation—he did not ever want to hear the word 'breeding' mentioned—was achieved long before the war.

During one of my many visits to his farm he buried twelve hermetically sealed alabaster vessels in the ground in the shape of a cross; they contained different seeds in the white of hens' eggs. He asked his son, who was 12 at the time, and other people to think of it at the time when it was buried. I cannot say anything about the results of this experiment but mention it here in order to show the rich variety that may arise when thinking comes alive and meditative work is done.

Hugo Erbe did not want to give lectures. He only liked to talk to individuals—serious, intimate talks between just two

Hugo Erbe

people, where words were often found to be inadequate in conveying one's meaning. I always felt I received many gifts on those visits and marvelled at this extraordinary friend and person who would seem rather like a magician to his friends.[17]

Martin Schmidt was someone I got to know after the war. He and the people on his farm had had to flee from the eastern part of Germany and coped with years of uncertainty as refugees most admirably. He and his wife had the strength, not only in those difficult times, to give warmth of heart to the people around them and speak to them of spiritual events. They had found their life's work in working to develop food that would feed the nation in accord with the spirit. We were able to meet quite often, as they were then living in Höri on Lake Constance, not far from 'my' children's village at Wahlwies. We would talk all night about the need to grow new food plants that had vital qualities.

Those night-time talks would usually be on the subject of how a breeder and grower might connect with the group souls of the plants he was working with and with the elemental spirits in the region, and how he might in this way stimulate and ask for true, stable mutations.

Martin Schmidt was working on the regeneration of local varieties rather than breeding new ones. He wanted to improve rye, the main bread grain, which he called the 'seismograph of cereal grains'. He would select particular grain sequences in an ear of rye, trying to discover the laws of the form-giving forces—successfully so and to great benefit. His method of counting out ears came to be widely known, the reward being his excellent 'Schmidt rye'.

Georg Wilhelm Schmidt continued his work in the Eifel, Ekkart Irion on his Grub farm, and so did others.[18]

We tried to create an aura with our breeding work at Sasterhausen that would help new plants to come into existence.

GETTING TO KNOW THE PIONEER PLANT BREEDERS 49

Martin Schmidt

Adalbert von Keyserlingk, 1938

We would then carefully observe them, assess them, and make further sowings. We knew that the dead need to have a hand in this and tried to include them in our thinking, feeling and doing.

We worked with the exercise Rudolf Steiner gave which is to look at a seed grain lying in front of one and perceive the flame that may develop from it as one meditates. We would collect ears that seemed suitable for further development, take them to the laboratory, and then begin our meditation work. It was not a matter of looking at a single grain but of finding the few in a small pile of about a hundred seeds where we had the impression that their aura was particularly luminous. We would sow them in the ground to connect them with the powers of the earth and the powers of those among the dead who were prepared to help us. Our concentration and meditation made it possible for the dead and for spiritual entities to connect with the work we were doing.

Transforming plant species and varieties

by Immanuel Voegele
(From Demeter, *monthly German journal of biodynamic farming practice, No. 12, 1938)*

In 1934 Privy Councillor Professor A. Bier published experimental findings on the transformation of yellow lupin into both blue and perennial lupin in the monthly journal *Der Züchter*, Vol. 6, No. 8, saying the following about these three lupin species in his introduction:

> All three have come to be widely cultivated here. They are distinct species with their own characteristics and cannot be confused. Even the two annual species, blue and yellow lupin, are far apart and stable as species. According to Hegi, varieties of a species can be crossbred but species themselves only with difficulty. Real experts on the subject of the Leguminosae whom I consulted told me that it is doubtful if even artificial hybridization would be possible. The perennial American species is, of course, even further apart from those two. The reader can realize, therefore, what it means when I transform one of these species, which after all are wild plants, into the other.

Following an observation he made in 1913, A. Bier felt it might be possible to transform one lupin species into another. He had been sowing yellow lupin on waste land to improve the soil. The plants ripened, self-seeded in autumn and then died off. The greater part of their seed that had dropped to the ground germinated the same year. The plants growing from them developed until killed by the first frosts. Another, smaller part went into delayed germination, that is, did not germinate in the autumn but only the following spring. The plants which then grew showed normal development, and when they were ripe the process from the year before repeated itself. When A. Bier found some blue lupins among those that had germinated in the spring, never having seen a blue one in the first year, he began to suspect that 'the delay in germination might have changed the yellow lupin seed so that blue ones suddenly appeared'.

Some years later the opportunity arose to take this up experimentally. In 1924 A. Bier heard of numerous yellow lupin seeds being turned up on the Charlottenhof estate (Mr von Klitzing) when a pine wood was partly felled and the soil ploughed. It was discovered that the last time the area had been sown with yellow lupins had been in 1869, before it became woodland. Special circumstances had delayed germination of the seeds that had dropped out, and the seeds remained in this condition for 55 years, until the ground was ploughed in 1924. A. Bier was able to obtain some of the seed for his experiments, and it was indisputably yellow lupin. It not only proved to be 100 per cent viable but also showed unusual energy in germination. The experiments were in many cases successful, in the direction A. Bier had thought they might go. One of the yellow lupins from the Charlottenhof seed gave rise to the garden lupin (*Lupinus polyphyllus*). Its descendants have all been perennial (no Mendelian changes).

On a second occasion one out of 100 Charlottenhof lupin seeds sown in 1933 showed retardation of growth. The first leaves were typical for yellow lupin, those that followed for perennial lupin. The pods, also showing the features of perennial lupin, produced 548 seeds. The plant died down in the autumn and did not come back the following year, thus proving to be a transitional form. Fifty plants were grown from the 548 seeds. All were definitely perennial. On another occasion, in 1934, three perennial lupins developed among plants grown from 2000 Charlottenhof lupin seeds that had all proved viable.

A. Bier reported a fourth case. In 1930, one of 37 fresh lupin seedlings grew taller than the others and had red cotyledons. The plant showed the habit of yellow and the leaves of blue lupin; it flowered yellow. Its pods were, however, found to contain blue lupin seeds. As F. Merkenschlager established, the pods themselves were those of yellow lupin. The offspring were all blue lupins (*Lupinus angustifolius*). The first generation was highly susceptible to disease. Subsequent generations produced healthy, strong blue lupins with all the characteristics of this plant. Another case where the same thing happened was noted the same year. In 1931, seed of Charlottenhof origin from the years 1926–29 were sown in a 400 square metre area. A. Bier wrote about the result as follows.

> The plants flowered in early July. Five blue-flowering plants flowered slightly before the yellow ones. The whole field looked rather strange, presenting a downright chaos of different types of lupins. Three stood out clearly—the *luteus*, the *angustifolius* and the intermediate type. They were clearly much influenced by external circumstances, with the first of them growing almost only in bare patches in the field, the second in lean and thirsty sites close to trees, the roots of which were depriving them of food and water. The intermediate type was scattered all over the field, as were the pure blue lupins.

A yellow-flowering plant with pods belonging definitely to blue lupin was also found in this field. Its descendants were blue lupins. Another yellow-flowering plant had otherwise all the characteristics of blue lupin, and one plant of intermediate type had characteristics of both yellow and blue lupin and concluded its growing phase by producing a rose-coloured flower. A. Bier also noted other, similar transformations in his experiments.

Apart from other things, it appears that Bier's experiments proved so successful because, as he tells us, the original Charlottenhof seed comes close to the Mediterranean wild form of *Lupinus luteus*. (Lupins were only introduced as a crop in Germany shortly before 1869, the year when the seed found in Charlottenhof entered into delayed germination.) Thanks to a number of favourable circumstances this seed, close to the wild state, was in delayed germination for more than half a century. Given the opportunity to germinate and develop, it then showed most unusual powers of transformation, producing blue and perennial lupins from yellow ones either directly or through transitional forms.

This clearly was a phenomenon of considerable import, but as Privy Councillor Bier wrote in a later publication ('Weitere Mitteilungen über Transmutationen', in *Der Züchter*, 1938, Vol. 10, No. 1), little interest was shown by the experts.

> One would have expected my experiments to have aroused the interest of scientists working in the field of genetics. Far from it, however. The objections raised are clear to me: the seed was taken for another ...

Bier refuted these and other objections, finally quoting Heraclitus in reply to the objection that it was improbable for yellow lupin to change into the American garden lupin: 'The improbable does often happen.' He added: 'Nature likes to hide herself.'

We know that nature will never allow herself to be limited in her creativity by narrow theories. It could never have produced the rich abundance of forms in all spheres of life if it had not been able to let the 'improbable' happen. Basically all new forms, all mutations that have ever appeared in organic nature were previously 'improbable'. It is certain that mutations and new developments in plant and animal evolution are special, exceptional cases, but it is equally certain that they always mark a step forward in evolution and that nature would not have its rich abundance of varieties and forms without such special cases. These are the creative exponents of her work and we have to say that, in spite of the 'arbitrariness', great wisdom lies behind their origin.

The natural world likes its creatures to be seen but prefers to hide the way they come into existence. It has put a veil over the 'how' of its activities. It seems to me that Bier lifted a corner of that veil when he got those remarkable results with lupins.

I myself saw the transformation of what were otherwise clearly distinct varieties of wheat, in many ways similar to the transformations Bier got with lupins. (I reported on this at the annual meeting of the national biodynamic farming association in 1935.)

In 1931, E. Pfeiffer let me have some glumaceous wheat seeds that had all the characteristics of spelt (*Triticum spelta*). These were fourth-generation descendants of a plant that showed few signs of being a cultivar when first found in Italy. The spikes and spikelets were insignificant; grains had developed, but were tiny and narrow. E. Pfeiffer started working with the plant and in just three years had a spelt that hardly differed at all in its external features from our white spelt. The rapid progression the plant made in becoming the cultivar would suggest that its ancestors had been cultivars that went wild for a time and because of favourable circumstances were able to continue as escapes.

In the years that followed I first of all sowed this seed alternately in August and December (see *Demeter*, 1930, No. 12, p. 241, 'Dynamische Wirkungen und ihre praktische Auswirkung'). I got normal descendants, similar to the parent plants. In 1934 I noted a plant among the December-sown plants with five ears that looked like a naked wheat, rather like *Triticum vulgare*. Thinking that an ordinary wheat seed had developed among Pfeiffer's spelt either because of carelessness or because an animal had dropped it there, I was about to tear the plant up. But the ear did not resemble those of the wheat varieties grown in the area and this made me hesitate. Taking a closer look I found that the ears showed distinct signs of transition from spelt to wheat. The spike was complete, though it was not as fully developed and firm as in ordinary wheat. The scales in the ears still reminded one of those of spelt in some respects. The plant was, of course, not pulled up but allowed to ripen. I harvested it myself. The transitional features showed very clearly when I removed the seed from the ear. On the one hand the spike was not yet entirely stable, and on the other it proved difficult to remove the scales from the hard, fully ripened seed. The seed was also elongated, looking very much like spelt seed.

The harvested seed was sown in two beds of 2 m^2 each at the end of September, and in another bed of the same size at the end of December the same year. The September sowing had tillered well before winter came. The December sowing emerged in February. It was soon thinned out so much by pheasants that only 13 plants remained. These grew to maturity. Until the ears developed, no appreciable differences could be seen between individual plants. I assumed that the daughter plants would be like their parents and that the intermediary form we had obtained in 1934 would remain stable through subsequent generations. This did not prove to be the case. When the ears had filled in July 1935 we found that against all expectation almost every plant had a different kind of ear. An observer who did not know where the seed had come from would have had to conclude that a mixture of all kinds of different spelt and wheat varieties had been sown.

On the one hand we had a wide variety of spelt ears. Some plants had extremely long, loose spelt ears, others short ones with the seeds unusually close together, and then there were all kinds of stages between these two. On the other hand—and in this case the variety was even greater—we had all kinds of wheat ears. One plant would have long, loose ears on long stems, another short, congested ears of any variety. You could not have had greater variety if you had taken each of the seed grains from another spelt or wheat variety. I had personally harvested the parent plant, removed the seeds, kept these in a covered container until sowing time and then sown them myself and so there could be no doubt but that every plant in that profusion of varieties had come from the same parent.

It might be thought that wind-borne wheat pollen had pollinated a spelt flower in 1934 and that this produced the intermediate form in the first generation which then divided up further in the second generation. On the other hand both wheat and spelt are self-pollinating and cross-pollination within the same species (without human help) is extremely rare. To my knowledge, natural cross-pollination between spelt and wheat has never yet been observed. If we nevertheless assume the improbable to have happened, with a spelt flower wind-pollinated by wheat pollen, it still cannot be explained, on the basis of established laws of heredity, why the daughter generation could produce so many different forms of spelt and wheat, and, what is more, how varieties could develop that came close to existing forms when none of these were found anywhere near the site nor further away from it.

I know that 'speltoid' mutations have been noted on occasion. In that case wheat had converted to spelt though there can have been no cross-fertilization with spelt. In the present case the opposite had happened and logically speaking it cannot have been regression. If the speltoid theory applied, one would have to assume a repetition of naked wheat developing from spelt.

Irrespective of all theory, the fact is that in a stroke of genius nature succeeded in letting a number of different wheat forms develop from a spelt. These forms have remained constant for four generations.

Privy Councillor Bier, having come to see the Pilgramshain spelt mutation in 1935 before it was harvested, gave it as his opinion that this was a phenomenon parallel to the transmutation he had seen with lupins in 1931 when yellow lupins (second Charlottenhof generation) produced a great mixture of different lupin types. A. Bier characterized his findings as follows, to quote his words again: 'The whole field looked rather strange, presenting a downright chaos of different types of lupins.' The same applied to the 1935 Pilgramshain spelt mutation, appropriate to the given case, of course.

E. Pfeiffer reported that a line of the original seed from which the spelt grown in Pilgramshain had come also produced a mutation when grown on in Holland. The phenomena seen there agreed with those described above in almost every detail. Bier's experimental findings were thus confirmed almost immediately afterwards in two independent cases. It is therefore reasonable to assume that Bier's work met with no interest among the experts because his findings do not fit in with accepted ideas. It is of course easier to assume the experimenter has made a mistake than to review an established theory or extend it. If the theory were extended, this would of course mean accepting Bier's findings.

When Darwin faced his doubts concerning the persistence of form in genera and species, he based himself on the observation that all external features of plants are variable. He made every effort to investigate that variability in detail and describe it, for he thought those variable features expressed the whole essential nature of the plant. Goethe, considered a precursor of Darwin, started his studies with similar observations. But he did not stop when he had established that all external characteristics of a plant are variable, but went on to look for the constant principle in the plant that was the foundation of the variable external features. Rudolf Steiner, editor of Goethe's scientific works (Kürschners National-Literatur edition) wrote the following concerning the Goethean and the Darwinian approaches:

> Whilst Darwin considered those features to be the whole of the organism's essential nature, and the conclusion he drew from the variability was that there is nothing constant in the life of the plant, Goethe went deeper, drawing the conclusion that if those features were not constant then there had to be constancy in something else, something that is the foundation of those external variations. He sought to evolve the latter, whilst Darwin's aim was to investigate and present the causes of that variability in detail. Both approaches are necessary and they are complementary. It is quite wrong to consider Goethe's greatness in organic science to lie in the fact that he was merely Darwin's precursor. His approach was much wider; it had two aspects: 1) the type, i.e. the laws that came to revelation in the organism—the animal nature in the animal, and life arising from life and having the power and ability to develop its inherent potential in many different outer forms (genera and species); 2) interaction between organism and inorganic nature and between organisms (adaptation and struggle for existence). Darwin only developed the latter aspect of organic life. We therefore cannot say that Darwin's theory was a further development of Goethe's original ideas, for in it one aspect of these has been developed further. It only considers the factors that make the world of life forms develop in a particular way and not the 'something' which is influenced and determined by those factors. If only one aspect is considered, this certainly will not lead to a complete theory of organisms ...

Goethe came to realize that the constant principle in the plant is something that cannot be perceived with the senses but is an inner power which, in spite of being ideal by nature, influences and determines visible realities. Soil, moisture, air—they provide the plant with the material it needs to develop. Light and heat lead to germination and growth. But the relative proportions in which substances are taken up and combined, their particular composition, and the form and configuration of the plant structures that develop are not determined by the substances, nor by the influences of moisture and heat, but by an invisible something in the plant itself. To the unbiased observer, plant growth is indeed the result of interaction between factors we can perceive with the senses and an invisible factor that comes from the living plant. Goethe called this the 'inner principle'. By taking it into account, he gained a wider and more comprehensive basis for his approach.

Goethe saw this inner principle take effect in as real a way as any factor that can be perceived with the senses. He found that the plant reveals its true nature in developing its leaves, and that every plant is made up of numerous leaf developments that go through a number of different metamorphoses. He summed it all up in the familiar words: 'In its forward development a plant is always all leaf'.

We can see the different stages of leaf metamorphosis in a lupin plant (cotyledon, foliage, petal, etc.). It is not through environmental factors but through 'life arising from life' that the lupin plant seeks to reveal itself in its ability to metamorphose the cotyledon into foliage leaf, sepal, petal, stamen and ovary.

The genus includes a number of species, including yellow, blue, white and perennial lupin. As species and varieties have the same ideal relationship to the genus as individual leaf forms have to the whole plant, we can say, on the basis of Goethean observation as we see a yellow lupin transformed into a blue or perennial one: 'Just as the inner principle relating to a single lupin is able to produce a range of metamorphoses of the lupin leaf if given the right environmental conditions, so is the inner principle belonging to the lupin genus or species able to let the plant come alive in different varieties and species.' It would go against organic nature if the plant type, which is higher than the individual plant, did not have the power to metamorphose.

In the light of extensive knowledge gained in genetics it may be assumed that the mutations described above followed changes in the seed's chromosomes or genes. (No relevant work has yet been done on lupins or spelt.) But anyone wanting to see such chromosomal changes as determining the transformation that follows would only have gone half the distance in his thinking. If I know from earlier observations, for example, that ruts going in a particular direction were made by a cart, I cannot stop at the horse and cart in looking for the factor responsible for those ruts. It should be evident to everybody that I have to go on to the driver who put the horses before that cart and decided which way it should go. Yes, the cart and the horses pulling it produced those ruts, but the determining factor was the driver's will.

August Bier's answer to the objection that there must have been a change in chromosomes first was both brief and exhaustive (*Der Züchter*, Vol. 6, No. 8):

> Let us assume such a chromosome mutation were found. This would not really explain anything. The question as to why plants change would simply have become the question as to why do chromosomes change.

Evading questions is not the attitude of a true scientist. If one tackles them one comes to the factor in organic development that Goethe called the inner principle, 'entelechy' or the 'self-determining principle'. Anyone who thinks this inner principle has no reality because it cannot be perceived with the senses—that is, anyone who believes purely materialistic observation to be all that is needed in the organic sphere—fails to take account of extremely important factors in the development of organisms. In that case it is not surprising to find that the insights gained are not enough to explain some phenomena.

Bier has undoubtedly opened up new avenues for plant breeding practice with his experiments, a perfect pioneering effort worthy of his achievements in medicine and forestry. It would be beyond comprehension if this work failed to gain recognition because it calls for a review and broadening of the views held so far in organics, not explicitly so but certainly implicitly. There is need for such broadening because at an earlier time Goethe's approach, which has its very roots in German cultural life, was superseded by Darwin's which came later and is more limited.

Koberwitz remembered

(Lecture given by A. von Keyserlingk in Driebergen, Holland, at St John's-tide 1987)

After the First World War my father was managing 18 combined family estates, a total of 7000 hectares. He had noted even then that more and more chemical fertilizer was needed year by year, yet the sugar content of the beet was going down all the time. So he asked Rudolf Steiner to help him with the problem of keeping the soil, plants and animals in good health. It is questions like these, arising from everyday practice, that led to the agriculture course in 1924.

Let me say in advance that my father revered someone who gave him the strength to cope with the many obstacles he had to overcome, some of them even before the course was held and many more afterwards. This was William of Orange. And so I am delighted to be able to speak in Holland, for from my early childhood days I have heard of the great sense, great personality and discretion of this man who made it possible for a nation truly to become a nation, in the face of opposition from Spain and France who were then great powers. With this ideal in his heart he was able to make it possible for the course to be held and the new approach to agriculture to become established, in spite of massive opposition from the chemical industry.

It took many negotiations until Rudolf Steiner finally came to the Koberwitz estate in Silesia at Whitsun 1924. For me, it was a tremendous event to have him there for those days, for I knew him already from the Stuttgart Waldorf School and revered him. Now, many years later, I realize that my parents actually only moved to Koberwitz for this purpose. The large manor house could accommodate and feed 100 people and was a place where an event could take place on such a large yet personal scale. What happened then was that for those who had the feeling for it, the 'gates of heaven' opened when Rudolf Steiner spoke and a wisdom streamed down that was not only for the head, the mind, but also for the heart. And because of this, my friends, this 'Koberwitz impulse', this new biodynamic farming method, is no 'recipe book' but something which can be used only on the basis of practical experience.

It was a glorious Whit Sunday. A pleasant breakfast was followed by a walk in the park. I remember above all the talk about the high iron content of the soil, so high that we could not drink the water and not even use it for washing; it had to be filtered. The filter plant was in a special room and Rudolf Steiner wanted to examine the water. And now let me tell you, the group was large, with people in their best clothes, most of the ladies in white and wearing white shoes; Eliza von Moltke was among them. When Rudolf Steiner got there one of the gardeners turned a tap and the rust-red liquid came out. I was of course standing close to Rudolf Steiner and was able to see every detail. When the dirty liquid came out the ladies started to shout: 'Dr Steiner, Dr Steiner, you'll get your feet wet. Please step back.' But Rudolf Steiner pretended not to hear, calmly stepped on to the ledge under which the liquid was draining away and asked for a glass. He filled it, held it against the sunlight and, cautiously, took a sip. Again the ladies started to fuss: 'Dr Steiner, please don't drink the water! You'll poison yourself!' But he drank some, nevertheless. He then reached into the glass with two fingers, rubbed them together and said: 'Pity, not enough arsenic. Otherwise it might have had a medicinal use.'

After the walk he withdrew with some anthroposophical friends and gave an esoteric lesson. This has not been published. He said important things about making the new farming methods known. He spoke of the significance of the farm as an individual entity, the spirit of the farm, and that the people who came together to work on the farm—farmers or also special teachers—should foregather once a day to be a spiritual vessel ready to receive the spirits that wanted to help us human beings. He was insistent that such a community had to be created to make it possible for the farm entity to develop. For, he said, the spirit world takes an interest in us! But the spirits cannot fructify the earth without us and we cannot move forward unless such a gateway is created. Rudolf Steiner spoke of the farm becoming an individual spirit if a community of people created such a vessel. A vessel through which the group souls of plants and animals and the elemental spirits who are called on so powerfully three times in the *Foundation Stone Meditation* are able to bring their influence to bear everywhere—on the climate, on fertility and in the community itself.

Apart from the farming methods, this was Rudolf Steiner's important social suggestion. For the new agriculture is not the egotistical concern of the landowner or manager, but a spiritual movement supported by people who want the earth to continue to give life.

Today, two generations later, it is much more evident that the earth is being destroyed on a global scale. We can see that unless we find ways of getting the spiritual world to help us we shall fall more and more into sickness, finally to perish together with our earth.

On the last day Rudolf Steiner gave a second esoteric lesson, which was at the request of the young people who were there. They came to Koberwitz from Breslau at 7 in the morning, for there was no other time available. The lesson has been published.[19] In it, Rudolf Steiner said that we young people were called upon—before that he had spoken of the earth's diaphragm—to seek Michael's sword on the altar which is below ground, find it and take it up into the world, the sword made of iron to defend the earth and prepare the way for Michael.

Those were Rudolf Steiner's words to people who wanted to work with the soil and to bring a dying earth back to life again.

Permit me to speak of a third, entirely personal experience that also happened in Koberwitz.

Rudolf Steiner had come from Munich where an attempt had been made to assassinate him. We young people from the estate had decided to protect him. One of us was always to be on guard, day and night. And so I, too, was on duty one day, from 8 to 10 in the morning, and was sitting outside the door of the room where he was breakfasting with his wife. As I sat there I thought: 'People are always asking him so many questions, perhaps I, too, should ask him something.' My cousin Aki came by. It was he who had finally got Rudolf Steiner's firm promise to come at Whitsun. He was another one who wanted to take 'the sword of Michael into the world'. As a young Russian army officer—he was a Balt—he had been taken to be shot seven times by the Red Army and had been saved each time—a strange destiny! I told him what I thought and he said: 'If you don't do it right now, seeing you are on guard, you'll not have another chance!' So I sat there and wrote down 30 questions—about the relationship of Waldorf pupils to one another and to their teachers and parents—a young person's questions. When Dr Steiner came through the door to go down the stairs to the lecture room I had to walk two metres behind him all the time, my hand on the pistol in my trouser pocket. You know, my friends, it took courage to address him, this man who belonged to the world! I took heart and said: 'Dr Steiner, I'd like to speak to you.'

He stopped and asked me: 'Yes, what is it you want?' It was something I'll never forget. He looked at me with great kindness and at that moment my soul life changed, for he was no longer the great head of our school, the famous speaker, the politician—all that dropped away. As he looked into my eyes he became someone I was able to speak to quite openly. Love spoke from his eyes and I became free, a free individual, as though we were equals, but full of respect.

Now I should have told him that I had 30 questions, but I only managed to ask one thing, and this was something I had not written down beforehand: 'Dr Steiner, I would be grateful if you'd give me a meditation.' He said: 'I'll be happy to give this to you, but right now I have to welcome our guests. Come and see me during the interval.' He then joined my parents at the entrance, welcoming everyone who arrived by shaking hands, all those people who had hastened from the station to the manor to attend the course.

Coffee and tea, sandwiches and frankfurters had been prepared for the many people during the interval. When Rudolf Steiner saw the large containers with the hot frankfurters he said: 'Oh, are the anthroposophists going to eat all those?' And he was delighted when they had vanished in no time at all.

I reported to him, standing by the square pillar in the entrance hall, saying: 'Here I am.' And again something happened. We were among all those people who were walking up and down the wide staircase, saying hello to each other, talking, drinking coffee—and yet we were quite alone! Something like an enclosed space had formed around us, as if in a temple. Rudolf Steiner said: 'I'll be happy to give you a meditation, but the first thing you must do, of course, is the review. The review is the basis of all meditation.' At the time I had as yet no idea how difficult that can be.

He then asked me to work with the Gospel of John,

pointing out that there are four sun aspects—the sun that sets at night, and another, spiritual Sun which then rises. Then again the physical sun that rises in the morning, when the spiritual Sun is setting.

It is really important for farmers who go to milk their cows in the morning or to mow a meadow to feel: the sun is rising up there in the sky, but the spiritual Sun which is now setting is taking with it everything I have been thinking during the night and will bring it back again. These things were still generally known in earlier civilizations—that the human being gives his questions to the Sun and after a time, one or several days later, receives an answer. I have also written about this in my book on Corsica.

If Rudolf Steiner gave you a meditation, that was a responsibility! The review, which I have now done every night since 1924, has become the foundation for my understanding of the Holy Saturday experience. It is the experience of entombment in the earth, the experience of Joseph of Arimathaea. Every human being goes through a review of his life after he has died (purgatory, in the terminology used by the Churches). It rids him of the bad things he had done in the past life and of his sins of omission. We know that Joseph of Arimathaea and Nicodemus took down the body of Christ from the cross and put it into Joseph's own new tomb. There Christ Jesus also had to go through a review—a review of the whole of earth evolution up to that time. He experienced the 'descent into hell' on that Holy Saturday, going through all the levels down to the centre of the earth where he put down the seed for a new future, to prepare for the Jupiter stage.

The earth is ageing and because of this it is getting harder and harder for people to gain access to this Holy Saturday experience. There can, however, be no resurrection without death and the 'descent into hell'. Nor will we be free to take up new things unless we have first got rid of the deeds done from cowardice, laziness or with evil intent in a life. This was what Rudolf Steiner meant when he advised me to practise the review.

I later had opportunity to go through an intensive review when the Nazis imprisoned me, 'the most dangerous anthroposophist in Silesia'. The SS took me to Breslau where I was on remand. People were quite decent to me—except for the interrogations—but I never knew if they might not take me to a concentration camp. So there I was sitting in my cell, with a lot of time in hand. I thought: 'What will you do with yourself all day long?' And I remembered that Rudolf Steiner had asked me to do the review. I began to review my life. And you see, there you do come to your own deep, dark places, and go through your own kind of hell.

First of all you discover all the things you have done wrong. But the things one omitted to do, or only did superficially, for show, are much worse—if you've given something to a beggar only so that you would be seen doing it, or if as a farmer you forgot to close the stable door at night, or if after a long day, seeing many difficult patients, you did not make that last house call. To live through this again is much worse than if one had slapped someone's face in a fit of anger!

Going back through life you finally see yourself as a young person again, a child, an infant. Then you come to your mother, feel yourself being born of your mother. Then one has to turn around, as it were, inwardly, in meditation, and one comes to the maternal ground and origin, to the element which at all times was known as Isis, or Persephone, or Madonna. All nations and civilizations knew this 'great mother' and venerated her.

Travelling in Monte Gargano I once found a steep rocky valley, quite bare, that was full of hermitages, each in isolation clinging to the rock walls high above. Hermits living at the 'abyss of existence' would go through this review of life there in preparation for further initiation. They, too, would come to the 'mother'. They would find her first in their own hearts, then as a wonderful painting in the convent church. We saw this Madonna, which is said to be by Luke the evangelist and was brought there for safety at the time of the iconoclasts. We experienced this Madonna as a guide to Michael, from the ground and origin to his great strength that can bring renewal for ourselves and for the whole earth.

You see, all this has been much on my mind in recent days. For not only is it 9 times 7 years since this impulse was given for the renewal of the earth and for new community building on the farms, but it is also only 12 years now to the end of the millennium. Anyone looking back on recent developments can easily see how short these 12 years that remain really are.

Many things have been prophesied for the end of the second millennium. My friend Dr Bühler has studied the prophecies from several centuries and recently given a lecture on this in Pforzheim in Germany. On that same day—it was the date on which the Koberwitz course had started—I gave a lecture in Öschelbronn, Germany, on the significance of the impulse we were given at that time. At the same hour when I was speaking about our tasks for the earth, Bühler said that every seer who was to be taken seriously had seen the earth grow dark at the end of this century, with the sun extinguished, and said one may assume that it will then no longer be possible to generate the energy for our closely interlinked industries. He said that a solar eclipse was forecast for 1998 [sic], the year when it is said 'the beast will be released from the abyss'. The number of the beast is said to be 666 in the Bible. Three times that number is 1998.

Such situations harbour the future, and we are already in the midst of it—a future created by human beings which we can and shall only meet in the Michaelic sense, iron sword in hand, without fear. For fear is the weapon of Ahriman who seeks to prevent us from experiencing the reality of the spirit and from learning from the review, which is the precondition for experiencing the Christ.

This, my friends, is what was meant by the initiations in the mysteries, by the hermits in the valley at Pulsano, and also by Rudolf Steiner when he asked us to seek 'the sword of Michael' beneath the earth's surface, find it and bring it up into the world.

He gave us meditations for this. You may know the lectures Rudolf Steiner gave in Pforzheim in 1914 on the pre-earthly deeds of the Christ. He spoke of the things we can do so that we do not perish as well, but have a part in the events that are to come. Read it up![20]

I would like to show you how the impulses that went out from Koberwitz took effect. For they work differently from the results obtained in conventional science. In modern science one must eliminate all subjective factors if one wants to obtain valid results. But it is not like this with the practical suggestions made by Rudolf Steiner. It is important for farmers and physicians to gain personal insight and experience of the way in which a preparation works.

I'd like to explain this by taking the stag bladder as an example. Imagine a stag, with tall antlers rising up high, with large, sensitive eyes, a very nervous creature. His sense organs are directed to the outside world. The antlers, shed every year and growing again in the spring, have developed out of the blood. This streams up, goes to the outside in the antlers, and thus perceives the sky and the world of stars. The blood makes the eurythmic A-gesture, as it were, in his antlers in order to experience the outside world.

Now compare this to a bull. His horns have developed from the skin and convey to the animal what is going on inside it, to regulate its metabolic functions. The stag on the other hand takes in cosmic powers, taking them into his organism through his eyes and antlers all the way to the kidneys. There they are turned round, as at the centre of a lemniscate, in a eurythmic E. Finally the powers of heaven, now coming from the kidney, are received into the bladder in a eurythmic O. I can only mention this briefly here, you have to study it for yourselves, this whole process which finally takes effect in the bladder.

Now you see, that is why we farmers take a stag's bladder and put yarrow flowers into it, the flowers of this miraculous plant with umbels that are also able to take up cosmic principles and bring them down to earth. Dr Steiner said that in yarrow spirits are wetting their finger with the finest sulphur to allow them to influence the world of nature.

As physicians we use yarrow to treat liver conditions. When a liver swells up, wanting to flood everything around it, we give yarrow preparations. On the farm we put the flowers into a stag's bladder, expose this to the sun in summer and bury it in the soil for the winter. When it is dug up in the spring, the farmer has a preparation which is first of all put into the compost heap where it has a regulating action, like a glandular organ in our metabolism.

This is how we must not only know the preparations Rudolf Steiner has given us but let them become living experience. They cannot be used like a cookery recipe. And if one were to try and make a chemical analysis of them, this would get us nowhere at all. Cosmic powers cannot be detected by such means. But if one experiences how the A changes to E and then to O, as those powers go through the antlers and kidney to the bladder, one will be able to use such a preparation rightly and it will prove to be a blessing.

We have to understand the things Rudolf Steiner has given us and truly use them. And so I would like in conclusion to come back once more to the darknesses at the end of this millennium, some of which are almost tangibly before us now. In this situation, Rudolf Steiner said, we anthroposophists need to be awake so that we will ignite the right light in us in good time. He spoke of this in the lecture on 7 March 1914 in Pforzheim, where he considered the beginning of the Gospel of John, transformed for our present-day conscious awareness. He transformed the words of the Prologue five times. First:

> In the beginning was the Word—and the Word was with God—and a God was the Word.[21]

This Word came to early man from the spiritual Sun and he was still able to understand it. Then came a time, in about the seventh century BC, when human thinking began to develop. It was not to be fully developed until the fifteenth century. Archaeologists call that time the Iron Age for it was a time when man also began to work iron.

Rudolf Steiner's translation then continued like this:

> In the beginning is Thought—and the Thought is with God—and divine is the Thought.

And then, my friends, Rudolf Steiner went from the 'luminous Thought, the living Thought' to the ability to remember into which the Christ impulse must enter. He went on to say that just as man received the power to think in the Iron Age, so the Christ himself has today implanted a new organ for memory in him. With this organ of renewed memory human beings will be able to connect the earth once again with the world of the spirit whence all order comes.

He went on to say that we need to practise to develop this organ, that we have to do something ourselves, for instance by being serious about the principle connected with our remembering, with the review, the Holy Saturday experience, the maternal ground and origin and with finding the sword of Michael.

And so, my friends, let me here, in this hall of the Herrnhut Brotherhood, Moravian Brethren who like many others have sought to unite with the Father ground and the Son spirit, say this meditation for you, a meditation that was given for all.

> In the Beginning is He (the Christ)[22]
> And the remembrance lives on
> And divine is the remembrance
> And the remembrance is life
> And this life is the I of man
> Flowing within man himself.

The divine light created in such remembrance is life, and this life flows in the human being. And the Christ, the primal power of the Beginning, is once again flowing in him.

> Not he alone, the Christ in him.
> When he remembers life divine
> The Christ is there in his remembrances.
> And the Christ will shine out
> As radiant life of remembrance
> Into every darkness we have in the here and now.

Yes, to overcome the darkness of the spirit that is all around us! This is a possibility which is coming towards us. Rudolf Steiner then gave his last version:

> In the Beginning was the power of remembrance.
> The power of remembrance is to become divine
> And divine shall be the power of remembrance.
> Everything that arises in the I shall be such
> That it is something which has come
> From remembrance made wholly Christian and divine.
> In it shall be life.
> And the radiant light shall be in it
> That shines into the darkness of our present time

From a thinking that remembers,
And the darkness as it is in our present time
Shall comprehend (this is an activity, my friends!)
The light of remembrance that has become divine.

You see how transforming the Prologue to the Gospel of John can illuminate this darkness and overcome it.

If we work with the gift we have been given in the Prologue to the Gospel of John, the darkness will not be able to touch us, nor will the beast, and we shall also be ready, when the end of the millennium comes, to make an active link with the living light of the Christ. We shall then be able to help the Hierarchies, which are always with us, so that the earth may continue.

This, my friends, is what I wanted to say to you, for it is something I care about deeply.

Learning about scientific research at the first Waldorf school in Stuttgart

In the early 1920s I was able to learn about the work Mrs Kolisko was doing at her Biological Research Institute. This was in the school grounds, a little distance away from the 'red wall', where the hall is today. We all knew and loved her, because she was so involved in and enthusiastic about her work. Her husband was our school doctor and also taught us. He was very popular with the pupils because he was so generous with his cough elixir. We had remarkably many coughs at that time, for it tasted wonderful. We also admired other qualities in him, of course.

When someone in the class broke a bone we were able to see Dr Kolisko bandaging and treating the fracture in such a way that no plaster cast was needed. I had watched him carefully and in my later medical work also managed to do without plaster casts. My wife broke her ankle on one of our study trips in the Corsican maquis. We were far from the nearest village. Dr Kolisko's teaching came strongly to mind again. And I got the same good result. The same thing later, when an old lady had broken the neck of her femur. She was sick and could not be taken for an X-ray. The fracture healed so well that it could no longer be detected at a later time. Dr Kolisko had shown us how Arnica can be used to heal, for it will mobilize healing powers better than any other plant.

Mrs Kolisko's biology research and Rudolf Steiner's lively interest in her work made a very deep impression on me in those days. We, the pupils of the upper school, were able to see a completely new method being tried out at the time, which was capillary dynamolysis. Countless series of experiments were done to create images in form and colour as fluids rose in filter paper and were then dried, making the ether forces visible for everyone to see. With breathless interest we found that it was possible to make the etheric nature of a plant visible by using its sap, or that of an animal by using its blood or lymph. We discussed the prodigious possibilities there would be in many fields, especially medicine, thanks to this method, and felt very much involved.

Apart from capillary dynamolysis, Mrs Kolisko also worked with artificial crystallization—not the way Ehrenfried Pfeiffer was doing it, but with freely developing, formed-out crystals that could be produced anywhere, even without a laboratory. She had had a 6-metre deep pit dug in which to do crystallizations so that she might show that the etheric presents itself differently below and above ground. One day when she was down in the pit carbon dioxide had accumulated there and she was unable to breathe. She was on the point of fainting when a pupil found her there just in time.

The most wonderful occasion was when Rudolf Steiner came into our classroom, his face alight with joy, and said: 'Now at last we are able to demonstrate the etheric, because of Mrs Kolisko's work with very small entities, and can prove to anyone who wants to see that science can be taken further and can find its way out of the dead end of materialism!' I shall never forget the joy Rudolf Steiner radiated as he said these words.

He then told us that from the beginning of the fifteenth century people had tried to enter more and more deeply into the dead matter of both the macrocosm and the microcosm, though they had no real aim in this. This had led to the division between belief and knowledge, with the spirit banished to the realm of belief and all things physical to that of soulless knowledge. It ultimately caused many people to be torn apart at the very core of their humanity.

Now, however, a beginning had been made to connect things of the spirit—the etheric being the lowest form of the spiritual—again with physical matter. The conscience of the scientist must also be part of the process. We were able to understand this world situation, both from the matter itself, from the method and the goal, and from the sheer joy in Rudolf Steiner's eyes.

These things were to play a major role in many of our lives, for an 'event' had happened in Mrs Kolisko's institute. And in Rudolf Steiner we had seen an initiator who found his suggestions understood and brought to realization. He had shown a new way at a time when science had grown destructive.

Another event I'd like to mention here was the following. Mrs Kolisko told us that Rudolf Steiner had greatly surprised her by saying that whilst sunflowers always turned towards the sun, they were not dependent on the sun but on Jupiter and should really be called Jupiter flowers. She had tried to influence sunflowers by using a number of different potencies, and this had proved successful. The next summer we saw an impressive row of sunflowers in front of the institute. The effects of different tin potencies were clearly evident.

Our attention was drawn to the yellow colour of the flower and we were told that the golden yellow fields of rape and mustard were Jupiter fields, letting the actions of the planet Jupiter become apparent in their immediate environment, and also in human beings. If one then learns that rape oil and mustard stimulate liver functions, one can see the relationship between the plants on earth, the organs in our body, and the planets of our solar system. One thus comes to see a wholeness that can restore the lost harmony between the natural world and the 'world of work' created by human beings.

We young people were intensely aware of the 'leap over

the abyss' from conventional scientific methods of investigation, where everything is explored yet there is no real goal and no concern about the consequences, to this new beginning we could see emerge before our eyes.

This step, which needed courage, was not identical with Goethe's achievements in the nineteenth century. Rudolf Steiner took this step, basing himself on Goethe, but it was a step from 'Goetheanism' to a new science of nature and of the spirit. It was a deed which Rudolf Steiner has asked us all to perform.

As schoolboys and girls we saw this as a move from one great mind to another, from an earlier generation to a new one. We felt a responsibility for what lay ahead of us in the future, strongly so, though not yet entirely clear in our minds. Today I know that we need to develop much greater awareness of these relationships and apply them in science. Our awareness must venture beyond knowing only the physical world, no longer negating conscience, as has been done until now, but rather calling it awake!

The first step is to gain insight into the ether world. This will take us on to perceiving the astral world, on which we are increasingly more dependent, though we know it not.

We were given a great gift to take through life with us in those days, into a future that was still unknown and was determined by the development of the sciences.

Some correspondence with Miss Windeck

I would like to add some passages from my correspondence with Miss Windeck who worked with us, running an experimental garden on the Ammersee lake. Our letters were in many respects personal and so the passages quoted have to be fragmentary, but readers may be interested to see the problems we were interested in at the time.

18 December 1969

Dear Miss Windeck,

... I have now decided to deal with the whole backlog of mail, and found a note from you that goes a long way back where you want to know something about breeding work, above all in connection with Pfeiffer and Voegele. We had a group with Mrs Wundt, Dr Vreede, Miss Riese, Messrs Eckstein and Pfeiffer. I also had many talks with Mr Voegele, Professor Bier and yourself. And I remember a sketch my mother had made 10 minutes after a conversation between Mr Stegemann and Dr Steiner.

I often talked with Samweber about Rudolf Steiner's view that our cultivars were progressively falling into decadence and could no longer feed the human race. It would therefore be urgently necessary to breed new plant varieties that would enable human beings to be human still at the end of the century and not automatons. He spoke about this a number of times, not only in the agriculture course and during meals at Koberwitz but also when talking to the young people in my father's study, to the medical profession, the workers at the Goetheanum, in the esoteric lessons and in private. He spoke about it above all to Ernst Stegemann, insistently and expressing pain, telling him to work with oat grass and wild barley as starting material for breeding.

Rudolf Steiner also talked to Wolfgang Wachsmuth when the latter had returned from China, speaking of the Chinese potato, *Dioscorea batatas*, as a plant particularly well able to take in the powers of light in the East. These would be much needed by Central Europeans in the near future, coming to them in their food. He compared *Dioscorea* to potato, a plant that brings earth forces into the organism. The well-known curve comes from this talk, intending to show how the light goes down to the earth in the East and moves across Europe.

It appears that he spoke often and in detail about a new bread grain with Ehrenfried Pfeiffer. I think he said new grain should be developed from the *Triticum* species—twitch or wheat grass, one-grained wheat (einkorn), or spelt.

Concerning twitch or wheat grass, which is a perennial, of course, and produces rosettes with rhizomes and no grain seed, Rudolf Steiner said in the agriculture course that since we manage to find four-leaved clovers we should also be able to find sufficient wheat grass seed. This is probably the wild grass from which all wheat species have evolved, developing into different varieties as they went in all directions of the compass and were subject to different earth radiation. I once found a broad-eared twitch or wheat grass on a botanical walk in Silesia. A rare but known species that might well give us the basis for a broad-headed wheat.

We worked with one-grained wheat, which was known even to the early lake-dwellers and does not let go of its glume unless force is used. A tanning process has to be used.

We also talked about amelcorn (*T. dicoccum*), that handsome plant where two grains always lie together in the ear, two triangular seeds with a high gluten content. It can be used in the unripe state. Then there is spelt, with grains that are quite similar to those of ordinary wheat.

Ehrenfried Pfeiffer apparently also talked with Dr Steiner about new oil plants, for he was trying again and again to grow oil seed from *Plantago*, greater plantain, 'white man's footprints'. He was partly successful; the seeds were slightly larger and did yield a little oil.

It was clear to us that this work would take decades and could only be done by people who developed a real inner relationship, in heart and mind, to the individual plants. We were aware that we needed to turn to the world of the spirit, for we would not achieve much on our own without its help and assistance from the group souls.

Each of us felt he had to go his own way, though it was evident that we would only achieve something together, for the work needed to be done in collaboration with the group souls, and successes gained by individuals had to benefit everyone. And this was clearly apparent.

The wild grasses were the greatest problem. Each of us related in some degree to a particular species, one more to wheat grass, another to oat grass or another species, and the most courageous to wild barley. Some used long-established species or varieties such as amelcorn, spelt, perennial woodland rye or local varieties from areas not yet exposed to modern cultivation and fertilizing methods. Other plants needed to be adapted to the climate or geography, Chinese potato being an example.

We were all agreed that firmly fixed species such as spelt needed to be loosened up, released from the rigidity of given forms, taking them back to their origins if possible. Only then would it be possible to produce something new, a mutation. Most of us were more or less experts in the field and knew that crossbreeding only produced varieties that would in due course disappear again. Hybridization and artificial pollination produce an unreal ahrimanic world that will dissolve again after a longer or shorter period. (People speak of the 'rapid deterioration' of varieties today.)

Those methods would never have allowed us to do as Rudolf Steiner had asked, which was to create a new bread grain from wild grasses and original species. This is the same way of working as when one uses Goethe's moral fantasy to perceive the reality of the archetypal plant. Otherwise one only has fantastic soap bubbles that will burst one day, even if one has invested a considerable amount of time in them.

Plants can be influenced in many ways, but it is only rarely possible to change them in the direction of truthfulness, where they will then remain stable.

Books and papers on cross pollination methods fill whole

libraries today, and new genetic laws are discovered all the time. We, too, have to take account of these, but not at the beginning of our breeding work but only at the end. If we should succeed in producing a new form, it would be necessary to develop this and produce a financially viable cultivar that was in accord with those laws.

A body of knowledge exists on influencing the environment. Rudolf Steiner spoke of these methods as being highly effective in his agriculture course. Just think of planting nettles, horse-radish, sainfoin and yarrow in field margins, and of mushroom meadows. We have gained considerable experience in this field.

In accord with this we tried to influence plants by intercropping, regulating moisture or dryness, sowing in ridges or furrows, spraying with calcium or silicon, transplanting from Switzerland to Holland, etc. Some of this proved effective. Exposing seed and plants to rhythms would also sometimes influence a particular property, in most cases probably because the property had originally come from such methods.

Rudolf Steiner's agriculture course included ways of influencing breeding experiments much more powerfully by one-sided application of the preparations, either using the Silica and the Horn Manure preparations intensively and often rhythmically, or specifically with the other preparations at different times and in different varieties. We tried to enhance the silica action by using kieselguhr (diatomaceous earth), so that fodder beet would be longer, more tender and aromatic. One can also use Horn Manure preparation to make roots that go down deep more rounded, or to fill the containing forms in which the preparations are made with seed and leave them in the ground through the winter. This will give major changes in the plants, roots and seeds.

We worked with moon and planetary influences. These can be activated by letting them irradiate water, by using metals or other plants. This, too, goes back to a suggestion Rudolf Steiner made in Koberwitz. He advised us to apply finely dispersed lead to roses that would not flower because of the high iron content in the soil. He also said wood fires gave better heat and were healthier if the trees had been planted when planetary movements were waxing. I planted cereal seed in the hollow trunks of appropriate trees and this had a considerable effect on the habit of plants.

Peas and barley from Pharaonic tombs which people had given me actually germinated, which was a surprise very much like the one A. Bier had with the lupins. Hugo Erbe had a trained voice, which he practised all his life, and this had an influence, as did the hands of another person, and the trained powers of will and heart of yet another.

We all knew that we could influence plants and their growth by these methods, but would only get transformations if we also made the connection with the spiritual realms of nature that were all around us.

The main task was to be awake and aware, and also to accept the changes which the spirits from those other worlds could make in the plants we offered to them.

We had to know our plants really well, of course, so that we would actually notice it when anything new developed. I have made hundreds of drawings of seeds, seedlings and the ears of grasses and cereal plants, so that I might be truly awake to and aware of them. Plants I worked with in that way would always show some kind of change.

Every single ear had to be examined long before and of course during harvest. One had to see it as it was and get a feeling for what it represented and where it wanted to go. Did it have new potential, was it 'preparing' for something, or was it perhaps at the end of its possibilities? Before sowing, before selecting the seeds, the same questions had to be asked of the seed grain. What was more or less apparent in the ripened plants still lay hidden in those seeds. We would select 20 or 30 seeds from thousands. This would be one nodal point in the breeding process. We had to find the few among the many that offered hope for the future. Here we were doing in practice what one is asked to do in the seed exercise. It was only rarely that one would see the small blue flame, but one was usually reasonably certain that this was the right seed grain.

It was an exercise where one needed to perceive life and its environs by becoming aware of one's own inner responses. More or less as it says in *Theosophy*: the heart must become an organ of perception. A relationship then develops between our inner nature and the seed. You had to grow absolutely still inwardly, for the challenge was to find the few grains that 'had future'. If two people had been working on this, there was rarely any doubt between them as to which was the right seed grain. And so we developed a completely objective way of using the powers of the heart for our future work.

I would again and again have occasion to observe the progress made by our friends, and many times we would all feel that the road that had been chosen did not take us any further.

Wild barley, for instance, proved as recalcitrant as it looks, though the grains did grow considerably bigger. Rye would often revert again when taken to a rye-growing area.

Hugo Erbe produced a new variety, a spelt that could be thrashed, with large grain good for baking. He called it 'golden corn'.

The mutations Immanuel Voegele produced in his Pilgramshain garden were an absolute miracle, going against all the laws of heredity. No scientist was prepared to take note, however.

One or two years after the war the same thing happened again in Hugo Erbe's garden. Different varieties originated in one seed grain. That was another miracle, but no one took any notice. Erbe was actually suspected of the worst kind of machinations. It seems we no longer have time and also lack understanding in research teams. Otherwise the experts would take such things up immediately. The phenomena of delayed germination and of dividing up into all kinds of varieties are research issues that would appear to be of prime importance for feeding the world, and I do find it surprising that one does not hear anything any more about work being done in this area.

The next problem is how to obtain a single type out of all that variety. A breeder may be moved to despair by the many different kinds of ears, but it does mean we have the possibility of a hundred new beginnings.

2 February 1970

Dear Count Keyserlingk,

Thank you for the Wernstein report. There the phosphorus puzzle really shows itself. I well remember early lectures by Wachsmuth, how he kept focusing on this question.

For me, it was no more strange at the time than everything else. Then, after 1945, I would persistently refer to the marked distinction between the five solid compost preparations and the valerian one. I even preached this to Pastor Vettersen when we were having coffee. Shortly afterwards he came to see me in my experimental gardens to tell me that Rudolf Steiner had told the priests that salt was the bread of plants, phosphorus their wine! When I met him on another occasion I wanted to thank him again for those words. But then he did not want me to speak of it. I wrote to him afterwards that I could not accept this, and that it surely was time for people to know this statement made by Rudolf Steiner. You'll know more about the priests' need for silence.

I feel it is important, however, when we talk about phosphorus again—keeping the group small—to make people aware of this special position which phosphorus holds, and therefore also the valerian preparation.

I feel it is not good to let plants drop once meditative work has been done on them. Hugo Erbe had to abandon his truly handsome wild barley because he did not have the money.

Your Berg Tabor wheat also had to go. My Manitoba mutation from Countess Polzer was eaten by sparrows at Becker's place. He has been putting nets over everything since then.

If you want to publish your paper on cereal grains, I'd be happy to let you have pictures of the Rome mutation. When Mr Dreidax visited me at that time, which was shortly before his death, he was extremely pleased to hear that I still have the photographs. With the first and greatest mutation in Pilgramshain we saw every form of wheat, from ears contracted into a sphere to long ones with awns, but no sign of amelcorn or one-grained wheat. I cannot agree with you on this. I had the Jena one-grain wheat for a long time, beautiful grains, but it never budged. With the Rome mutation it was possible to get constant lines, but I did not get the beautiful form of the Künzel wheat. It was always dividing up again.

I remember you telling me that plants sometimes reach a kind of zero point and then quite suddenly something happens. Surely this is something you should also include in your paper?

24 February 1970

Dear Miss Windeck,

I am delighted to hear that one-grain wheat and amelcorn were not used on that occasion in Pilgramshain. One-grain wheat and spelt are plants I have also worked with. The seeds grew larger and larger, especially with spelt, and some did thresh out naked. The amelcorn did, of course, have excellent gluten; the grain was glossy, firmly attached to the glumes.

The business with the zero point is something I'd like to do more work on.

The difference between a genuine mutation and a variety is often difficult to establish, let alone describe, if one does not know the parents.

In any case, a mutation is stable, whilst a variety will metamorphose again and again.

I do not think a modern breeder supplied with our plants will make them economically viable. This is clearly something else that we'll have to do ourselves, otherwise too much of it will be ruined.

To come back again to the Schmidt rye. I fear it does divide up in some situations. But if we can strengthen it so that it will be stable and not degenerate after a few generations, Martin Schmidt can truly be said to have done something for the soul of rye.

I hope this will be possible, for he has done an immense amount of work and above all given much love to it.

A cycle came to an end with this correspondence, written almost 50 years after the work we all did on a bread grain for the future. I hope it may be taken up again some day, by young people, as we once were, who will move on into a new age, at least for Europe, with confidence.

I wrote the manuscript first of all with my wife, and later with two other faithful helpers. With chapters having to be revised and corrected several times and then copied again, work that could not be done under my direct supervision, it has perhaps not quite turned out the way I had intended, from the heart. Some more personal things, names that no one knows any more today and events that would need to be explained and would then take up too much space had to be left out. But I hope that it still speaks of the things that mattered to us then, will matter again today and probably even more so in the future.

Öschelbronn, January 1993

Photographs of the experiments done at Pilgramshain

Rome mutation by
E. Pfeiffer
Harvest 1935,
Pilgramshain

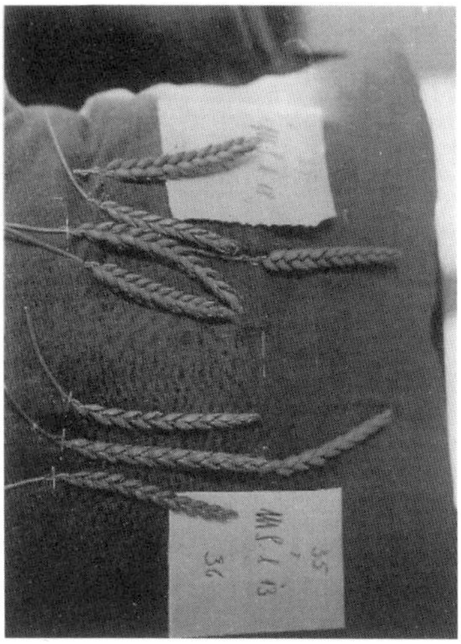

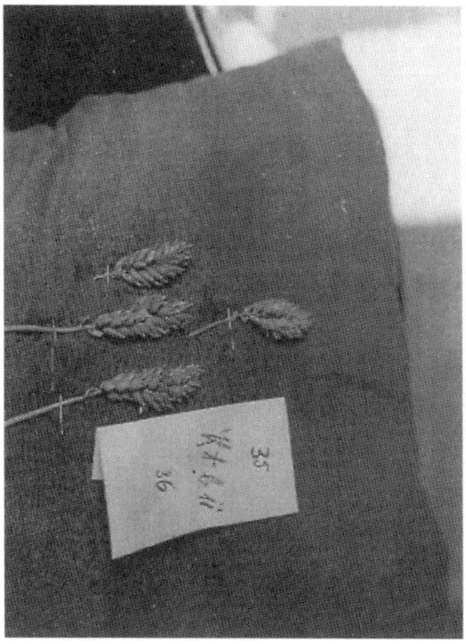

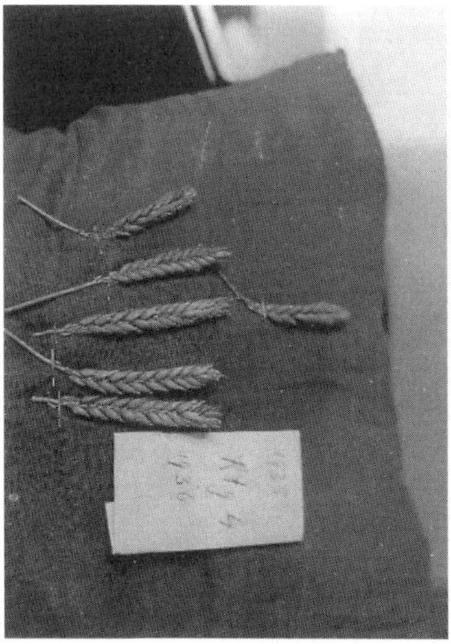

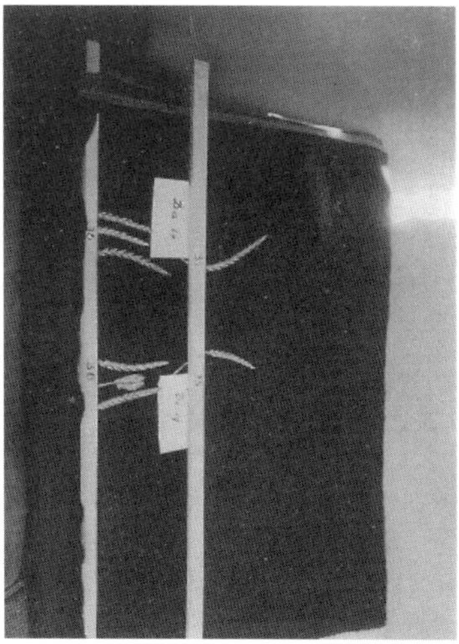

PHOTOGRAPHS OF THE EXPERIMENTS 91

*Mutating spelt
4 July 1935, Pilgramshain*

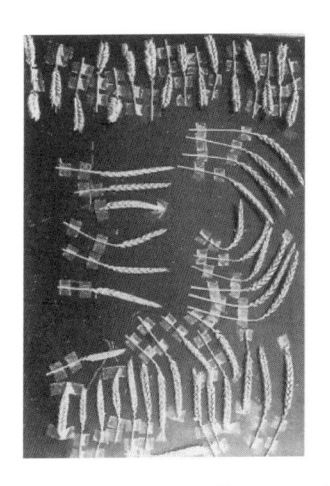

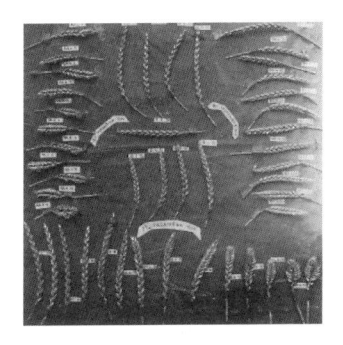

*Mutating spelt, collection of a wide variety of forms from 1935
to 1944, Pilgramshain*

Rome mutation by E. Pfeiffer, sowing in Pilgramshain in 1935

Complete set of forms

PHOTOGRAPHS OF THE EXPERIMENTS 93

Spelt mutation, C. Voegele, Pilgramshain

First sowing of mutated plant, end of 1935

94   DEVELOPING BIODYNAMIC AGRICULTURE

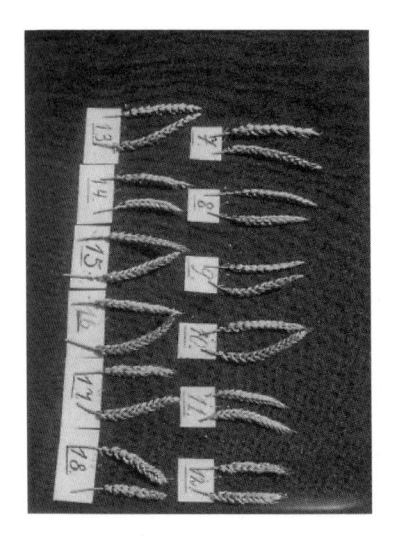

Rome mutation by E. Pfeiffer.
1939 harvest, Pilgramshain

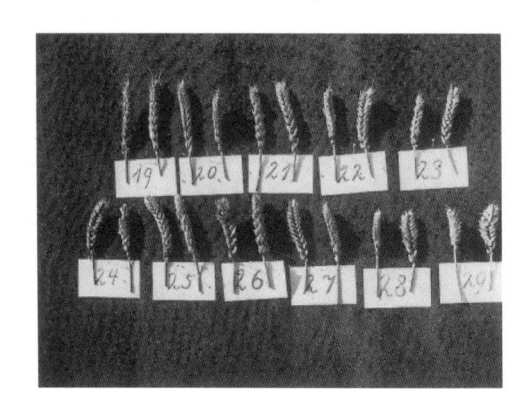

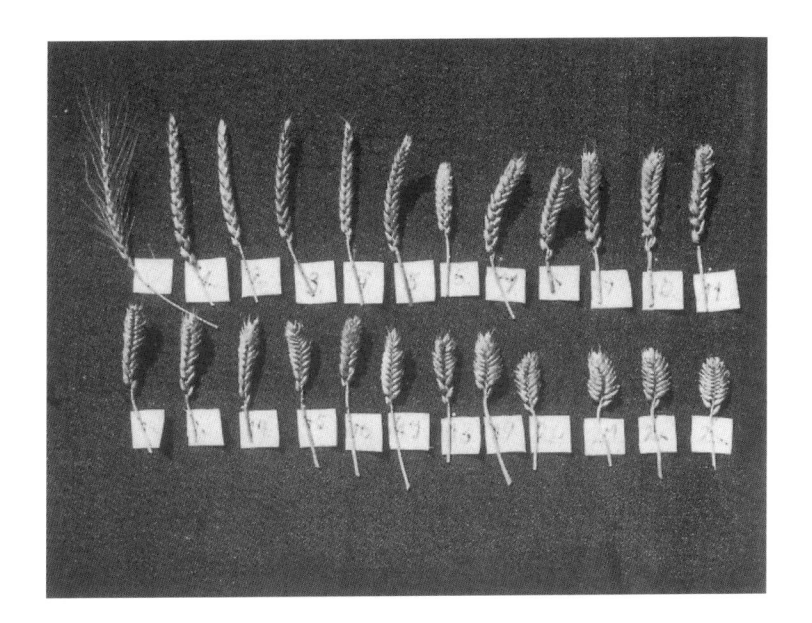

PHOTOGRAPHS OF THE EXPERIMENTS 95

Different forms of Rome mutation by E. Pfeiffer. 1937 harvest, Pilgramshain

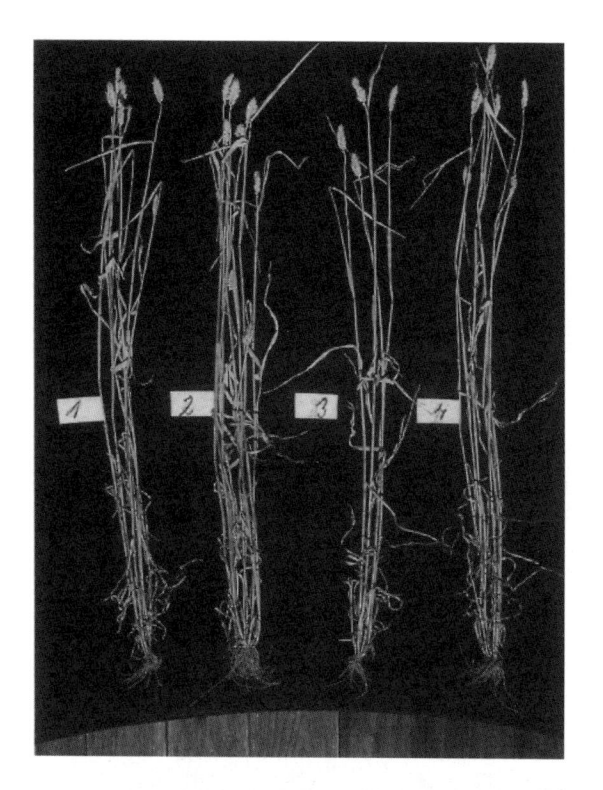

Rome mutation by E. Pfeiffer. 1939 harvest, Pilgramshain

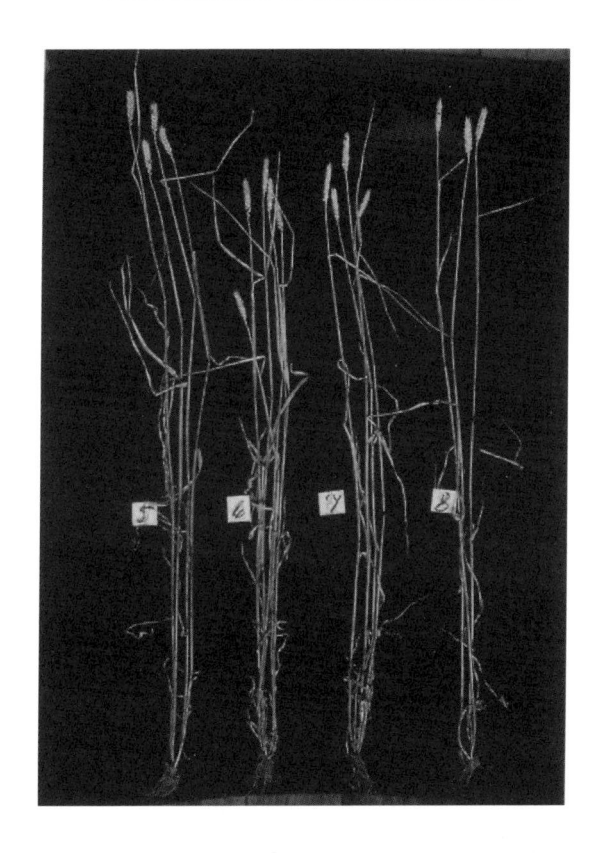

PHOTOGRAPHS OF THE EXPERIMENTS 97

98 DEVELOPING BIODYNAMIC AGRICULTURE

PHOTOGRAPHS OF THE EXPERIMENTS 99

The revitalization of the ether powers

by Erich Kirchner

The following article was originally published in the 1974 Weihnachtsrundbrief of Weleda AG, Arlesheim. It makes a fitting conclusion to this small work. The subject may seem far removed from plant breeding, but those who seek with their hearts will find that it brings a new note to the issues we have been considering.

We have been studying the preparations for Christianity in the post-Atlantean mysteries. Let us now try and get closer to the central event in human evolution. The focus shall be on the coming of the Christ spirit and above all on the Mystery of Golgotha—the death of Christ Jesus, when his blood flowed into the earth. Why is this event of such great significance to us? The flowing blood changed the earth, which was dying; the earth ether was transformed, just as the human etheric can be given new life.[23] The transformation of 'substance', which this brought with it, is the basis of our pharmaceutical work and it seems of fundamental importance to be fully aware of this.

In 1906 Rudolf Steiner said there should be food laboratories where new foods were produced, for instance a product that 'will be greater in quality than milk'. 'The time will come when students of spiritual science will be working with chemistry in harmony with evolving nature and not the natural world that has already evolved.'[24]

In the first medical lectures it was said that it was the task of the pharmaceutical laboratories that had been established in 1921 to watch plants patiently so as to learn their processes and use our knowledge of those processes to make medicines. Compound medicines based on medicinal plants and vegetabilized metals are examples of developments that have arisen from this. Their use has borne much fruit already.

In the agriculture course held in Koberwitz, a way was shown in which the plant and animal worlds can be maintained to provide foods, preventing their complete degeneration and extinction. It is in the nature of these things that the three tasks are difficult and do not easily gain recognition from the world. In 1906 no 'theosophist' ever thought of putting Rudolf Steiner's teaching into practice, for they would first look for their own way of higher development. The two later impulses came shortly before Rudolf Steiner's work on earth came to an end. There were people then who wanted to put his teaching into practice, but it is not easy to move on from present-day training in science to finding ways of bringing new life into the etheric. The only people who may perhaps have had what was needed for substance transformation were probably those who had been alchemists in medieval times and knew of the Rosicrucian quintessence.

Education, education for special needs, anything connected with the arts will be much more easily accepted in this world. When it comes to the new impulses to save the earth and human spiritual development, both of them connected with the right nutrition and the right medicines, the opposition is especially powerful. We can see this clearly in recent developments concerning drugs legislation, and also when it comes to artificial fertilizers and plant protectives.

Rudolf Steiner foresaw the ultimate consequences of the modern way of thinking only in terms of profits. He also knew that the way of thinking used in materialistic science would not help people make the connection with the healing powers needed for man and earth. If one considers only physical matter this does not take one beyond the sphere of action of an earth and a natural world that are hardening and falling into decay.

Why does the power that makes connection with those healing powers possible, lie in the blood that flowed on the cross?[25]

The earth as a physical and material creation must perish, and part of the etheric world will also die. There are ether powers that take us on into the future, and humanity needs to connect with these, which, however, will only be possible if we actively seek the way to the Christ. The blood flowing on the cross is the wellspring of the new creation that is to take us on to Jupiter. The new etheric substance comes from it through which the earth is transformed and human beings are given new powers of life.

What is the role of blood in the history of the human race, and what should the blood become through the event on Golgotha? Human blood is the instrument which enables the human I to influence the development of the body, the organs and the principles of life. Rudolf Steiner spoke about this from different points of view in many of his lectures. This activity began in Lemuria and continued to configure the human body in Atlantean times. The human form was still very different then. The human I, still in the keeping of the Hierarchies then, and the power of the human soul were working to develop the human form via the blood. Rudolf Steiner said in 1907: 'The effect on the blood in the early stages of human evolution was enormously powerful ... The sculpting, configuring power came from the I, via the blood, as the human body developed.'[26]

The hereditary powers based on the blood prepared individuals in later Atlantean and post-Atlantean times who were chosen to be initiated into the mysteries and become guides for humanity, sharing the responsibility for the future of humanity with the gods.[27]

Where do we find ways to come closer to this open secret? Where are the streams that have the knowledge of the blood mystery, preserve and guide it? Grail Christianity, the Templars and the Rosicrucians were the harbingers who prepared humanity for the age of the spiritual soul, serving those impulses. But like the whole of esoteric Christianity they worked more in secret, hidden ways, and have sometimes been persecuted and even destroyed—thus the Templars, for instance, who among other things had vowed that every drop of their blood should belong to the Christ. Other streams are more evident and we are able to discover them. They are connected with the Dominicans.

Studying the Dominican monasteries in Florence, Ita Wegman said to me in the late 1920s: 'The Medici, above all Cosimo and Lorenzo, took up the Rosicrucian and Templar impulses. That is where we must look for the background to their actions.' It was also in connection with this that Rudolf Steiner asked Maria Röschl to study the Christianity that was independent of Rome and how it moved from Greece and through the Goths to Florence.

At the beginning of the spiritual soul age, when the Renaissance was at its height in Florence, Cosimo de Medici rebuilt the San Marco Convent. He was a friend of the abbot's and had his own cell at the convent, where he would withdraw at times. He got the Dominican Monk Fra Angelico to come from a small monastery in Fiesole to San Marco, asking him to paint pictures in every cell and in the rooms used by the community. These are meditations on the life of Christ Jesus.

Entering the convent one first of all sees the painting at the end of the cloisters. It shows Christ Jesus on the cross, his blood flowing into the earth. Dominicus, the founder of the Order, is at the foot of the cross. As you walk on and visit the monk's cells, you find that more than half the paintings all show the Christ on the cross and his blood flowing into the earth. Different individuals may be included in the scene. In the Cosimo cell, for example, John, Mary, Peter the Martyr and Cosmos, the martyred physician, one of the Medici family's patron saints, stand by the cross. You will look in vain for a representation of the blood secrets at the other great Dominican convent in Florence. It has paintings showing the Church triumphing over heretics, the Dominicans as the Lord's dogs, Thomas triumphing over Averroës the Arab, and others.

So what is this image of the blood flowing on the cross? 'On the cross, the excessive egotism in human blood flowed from the wounds of Christ Jesus in a mystic, real way; it was given in sacrifice. If this blood had not flowed, egotism would have become ever greater in the human blood as evolution continued ... This deed of love has saved the human blood from its self-seeking.'[28]

Many paintings of Golgotha show a skull at the foot of the cross, with one of the rivulets of blood flowing into it. The deadest part of the earth is transformed by the blood. If you continue your investigations as to what lies behind the meditation paintings at San Marco, you discover the close connection that existed between the convent and the Platonic Academy which Cosimo de Medici founded at the beginning of the fifteenth century. Connections then emerge with Rosicrucians such as Pico della Mirandola (e.g. his work *De hominis dignitate oratio*), Agrippa von Nettesheim and others. They, too, knew the secret that man was to be the Tenth Hierarchy. This lived in the hearts and minds of many who were connected with the convent and the Platonic Academy in the fifteenth century, and Botticelli actually showed it in a painting, which caused him to come under censure from the Church.[29]

These are the impulses we find in Florence, the city of which Rudolf Steiner said that Rome-free Christianity lived. This was at the beginning of the spiritual soul age. Five hundred years have passed since then. We know from Rudolf Steiner that the earth is inexorably falling into decay and that this is necessary so that evolution may take it to the Jupiter stage. We can perceive this process today. The etheric of the earth and of man is also moving towards this process of hardening and of death. The great danger is that these processes go too fast, so that the goal of Earth evolution will not be achieved.

A new, spiritualized ether developed for the earth from the blood of Christ Jesus that flowed on the cross. It is important that we make a connection with this newly created spiritual and etheric 'substance', for then we shall be able to meet our responsibility for the earth. For our own, essentially human, etheric, which is also dying, we can do the same by making a connection with the Christ deed and by recognizing the Christ. In connection with the etherization of the blood we can fulfil the mission which Rudolf Steiner expects us to fulfil and which our karma demands.

The healing powers and also the medicines of the future will not depend on the effects of substances but on growing insight into how the ether powers are revitalized through the blood that flowed on the cross.

If we think on this in our present life, taking it into our minds but above all also into the will, we can be certain of developing faculties in a future life that are not yet possible at the present time.

Notes and References

1. Steiner, R., *Agriculture* (GA 327), lecture of 11 June 1924, tr. G. Adams, London: Bio-Dynamic Agriculture Association 1977. Alternative translation by C.E. Creeger & M. Gardner, Kimberton PA: Biodynamic Farming and Gardening Association 1993.
2. Alanus ab Insulis, *Der Anticlaudian*, translated and with an introduction by Wilhelm Rath, 2, Aufl. Stuttgart: J. CH. Mellinger 1983.
3. von Keyserlingk, A., *Und sie erstarrten in Stein. Frühe Mysterien in Korsika*, Basel: Die Pforte 1983.
4. Heinze, H., *Mensch und Erde*, S. 152, Dornach: Philosophisch-Anthroposophischer Verlag 1983. [Guldesmühle was a biodynamic estate with its own mill and bakery in the eastern part of the Swabian Alps.]
5. von Keyserlingk, J., *Erlöste Elemente*, 2, Aufl. Stuttgart: J. Ch. Mellinger 1983.
6. Meyer, R, *Wer war Rudolf Steiner?*, Stuttgart: Freies Geistesleben 1961.
7. See facsimile on page 12. (The German edition includes a facsimile of the entire note.)
8. *In the mornings*

steadfastness	left leg
certainty	right leg
love	left hand
hope	right hand
trust	head

Steadfastly I take my place on earth	concentrate on left leg
Certain as I walk through life	concentrate on right leg
Love in the core of my being	concentrate on left arm
Hope in all I do	concentrate on right arm
Trust in all my thinking	concentrate on the head

NOTES AND REFERENCES

> These five take me to my goal.
> These five brought me into earthly existence.

In the evenings
> Look back on the day's events,
> back to front, as far as you can in pictures.

9 Steiner, R., *Pre-Earthly Deeds of Christ* (from GA 152), lecture given in Pforzheim on 7 March 1914, tr. H. Collison, 2nd imp., North Vancouver: Steiner Book Centre 1976.
10 Steiner, R., *Balance in the World and Man* (from GA 158), lecture given in Dornach on 21 November 1914, tr. M. Adams, 2nd ed., North Vancouver: Steiner Book Centre 1977.
11 See page 51 ff in this volume.
12 Bier, A., *Der Wald in Sauen*, Schönau: Erde und Kosmos 1976.
13 Steiner, R., *The Etherisation of the Blood* (from GA 130), lecture given in Basel on 1 October 1911, tr. A. Freemann, D. Osmond, London: Rudolf Steiner Press 1971.
14 Quotes from the Bible were taken from the King James version or the translation by K. Bittleston, as appropriate.
15 Kolisko, E. & L., *Die Landwirtschaft der Zukunft*, deutsche Ausgabe 1953.
16 Selawry, A., *Ehrenfried Pfeiffer, Pionier spiritueller Forschung und Praxis*, Dornach: Verlag am Goetheanum 1987.
17 Erbe, H., 'Präparate zur Förderung des elementaren Kräftewirkens, Sonderdruck', *Erde und Kosmos* 1983, *Erde und Kosmos*, Sonderdrucke aus diversen Jahrgängen, Schönau.
18 von Wistinghausen, E., 'Gesetzmäßigkeiten beim Roggen, die Ährenbeetmethode von Martin Schmidt', *Elemente der Naturwissenschaft,* Heft 6, 1967, Dürnau: Verlag der Kooperative.
19 Steiner, R., *Youth's Search in Nature* (from GA 217a), lecture given in Koberwitz on 17 June 1924, tr. rev. by G. Karnow, A. Wulsin, Spring Valley: Mercury Press 1979.
20 Steiner, R., *Pre-Earthly Deeds of Christ* (from GA 152), lecture given in Pforzheim on 7 March 1914, tr. H. Collison, 2nd imp., North Vancouver: Steiner Book Centre 1976.

21 Steiner, R., *Verses and Meditations*, p. 175, tr. G. & M. Adams, London: Rudolf Steiner Press 1972.
22 In the 1964 edition of *Wahrspruchworte* it says:

In the Beginning was remembrance,
And the remembrance lives on ...

However, both Countess Keyserlingk's and Gottfried Husemann's notes have the version given here. In the 1935 edition of *Wahrspruchworte* it says:

In the Beginning lives He
And the remembrance lives on ...

23 Steiner, R., *The Etherisation of the Blood* (from GA 130), lecture given in Basel on 1 October 1911, tr. A. Freemann, D. Osmond, London: Rudolf Steiner Press 1971.
24 Steiner, R., *Ursprungsimpulse der Geisteswissenschaft* (GA 96), lecture given in Berlin on 22 October 1906.
25 Steiner, R., *The Fifth Gospel. From the Akashic Record* (GA 148), lecture given in Berlin on 10 February 1914, tr. A. Meuss, London: Rudolf Steiner Press 1995.
26 Steiner, R., *Ursprungsimpulse der Geisteswissenschaft* (GA 96), lectures given in Berlin on 25 March and 1 April 1907.
27 Steiner, R., *Esoterik und Weltgeschichte*, lecture given in Berlin on 28 October 1904.
28 See Note 26.
29 *The Assumption*, National Gallery, London.